Springer Tracts in Modern Physics
Volume 199

Springer
Berlin
Heidelberg
New York
Hong Kong
London
Milan
Paris
Tokyo

Springer Tracts in Modern Physics

Springer Tracts in Modern Physics provides comprehensive and critical reviews of topics of current interest in physics. The following fields are emphasized: elementary particle physics, solid-state physics, complex systems, and fundamental astrophysics.
Suitable reviews of other fields can also be accepted. The editors encourage prospective authors to correspond with them in advance of submitting an article. For reviews of topics belonging to the above mentioned fields, they should address the responsible editor, otherwise the managing editor.
See also springeronline.com

Martin Schmidbauer

X-Ray Diffuse Scattering from Self-Organized Mesoscopic Semiconductor Structures

With 100 Figures

Springer

Dr. Martin Schmidbauer
Humboldt-Universität zu Berlin
Institut für Physik
Röntgenbeugung an Schichtsystemen
Newtonstr. 15
12489 Berlin, Germany
E-mail: schmidbauer@physik.hu-berlin.de

Cataloging-in-Publication Data applied for

A catalog record for this book is available from the Library of Congress.

Bibliographic information published by Die Deutsche Bibliothek

Die Deutsche Bibliothek lists this publication in the Deutsche Nationalbibliografie; detailed bibliographic data is available in the Internet at http://dnb.ddb.de.

Physics and Astronomy Classification Scheme (PACS):
61.10.-i; 68.65.-k; 68.35.Ct; 81.16.Dn; 81.07.-b

ISSN print edition: 0081-3869
ISSN electronic edition: 1615-0430
ISBN 3-540-20179-3 Springer-Verlag Berlin Heidelberg New York

Springer-Verlag is a part of Springer Science+Business Media

springeronline.com

Typesetting: Author and LE-TeX GbR, Leipzig using a Springer LaTeX macro package
Production: LE-TeX Jelonek, Schmidt & Vöckler GbR, Leipzig
Cover concept: eStudio Calamar Steinen
Cover production: *design &production* GmbH, Heidelberg

Printed on acid-free paper SPIN: 10955062 56/3141/YL 5 4 3 2 1 0

Preface

This book represents a survey of the outstanding capabilities of X-ray diffuse scattering for the structural characterization of mesoscopic material systems. The term *mesoscopic* comprises length scales between a few and some hundreds of nanometers and is of particular importance for low-dimensional semiconductor nanostructures such as quantum dots and quantum wires. In the past decade these objects have attracted huge interest because they can be used to create "new" material systems with altered optical, electronic, and transport properties. In order to achieve this goal, the geometrical parameters of these objects, for example the shape, size, atomistic structure, and strain state have to be well known and controlled through modern fabrication techniques. Among the most favorable candidates are the "self-organized" growth techniques; however, many aspects of the growth mechanisms are still not fully understood. This leads to the necessity to comprehensively study the real structure of mesoscopic systems. Well-adapted characterization techniques are required which cover a wide interval of length scales, ranging from the atomic scale up to the micrometer range.

I review the particular advantages of X-ray diffuse scattering for the structural investigation of mesoscopic systems. After a general introduction to the theoretical and experimental aspects, selected examples are discussed that demonstrate in detail the capabilities of X-ray diffuse scattering. These examples originate mainly from research performed in the X-ray group of the Humboldt University in Berlin. They are supplemented by a few examples from other laboratories in Berlin, Grenoble, Kiel, Linz, and Munich.

I am indebted to many colleagues, without whom this book would surely not have been possible in the present form. I would like to express my particular thanks to Daniil Grigoriev, Michael Hanke, Rolf Köhler, Ricarda Opitz, Tobias Panzner, Helmut Raidt, Peter Schäfer, and Thomas Wiebach. Many results presented here were achieved as an output of their successful research in the X-ray group at the Humboldt University in Berlin. I have also greatly profited from the expertise of Sergey Stepanov and Elena Kondrashkina (APS, Argonne), who are among the world's leading experts in X-ray diffuse scattering from interface roughness. I had inspiring discussions with Hartmut Metzger (ESRF, Grenoble), Vaclav Holý (University of Brno), Julian Stangl (University of Linz), Vladimir Kaganer (Paul Drude

Institute, Berlin), and many others. I am also grateful to Herbert Wawra (Institut für Kristallzüchtung, Berlin), Fariba Hatami (Humboldt University Berlin), Rudolf Hey (Paul-Drude Institute, Berlin), and Roman Sellin, Lisa Steimetz, and M. Pristovsek (Technische Universität, Berlin). The X-ray scattering experiments were performed at synchrotron radiation facilities, where I received technical support from Frederico Zontone, Bernd Struth, Gavin Vaughan, Oleg Konovalov, and Alexei Souvorov (ESRF, Grenoble), as well as from Stanislav Bessedine, Holger Rhan, Oliver Seeck, Oliver Bunk, Wolfgang Drube, and Dimitry Novikov (HASYLAB/DESY, Hamburg). Finally, I am especially grateful to Jane Richter for valuable support in proof reading the manuscript.

Berlin, October 2003 *Martin Schmidbauer*

Contents

1 A Brief Introduction to the Topic

1.1 Mesoscopic Semiconductor Structures

During the past decade, advanced techniques of crystal growth have been developed to tailor the electronic properties of materials. This fascinating prospect of fabricating *new materials* that exhibit modified physical properties is based on the competition between the dimensions of the structure and the de Broglie wavelength of the electrons. In the case of semiconductors, which we shall focus on in this book, the exciton Bohr radius – which is typically a few tens of nanometers – plays a key role since it is related to the corresponding exciton binding energy and thus to the electronic and optical properties. As long as the geometrical extension is well above the exciton Bohr radius, the electronic structure is exclusively a material property. However, when the geometrical size is below the exciton Bohr radius – i.e. the exciton is confined to a nanometer-sized region inside the semiconductor – the electronic structure depends on the dimensions of the structure and can be substantially altered. Then, quantum size effects dominate.

In principle, the confinement can be present in one, two, or three dimensions. In the one-dimensional case of a quantum well, the thickness of a buried thin layer is below the exciton Bohr radius of the bulk material. In that case, the exciton size and, thus, the optical properties are very sensitive to local thickness fluctuations of the well. This interfacial roughness may even induce quasi-localized exciton states varying in binding energy, which give rise to inhomogeneous broadening of the area-integrated luminescence spectrum. These localized states are also present in quantum wires and play a key role in the optical and transport properties of disordered (rough) quantum systems. Roughness is therefore a crucial structural parameter that strongly influences the optical properties and needs detailed analysis.

The effects of localization are especially strong when the exciton is confined not only in one dimension, as in quantum wells, but in two (quantum wires) or even in all three dimensions (quantum dots). In the latter case, localized discrete states are induced and an atom-like luminescence spectrum is observed [1]. The optical properties of quantum dots can be exploited to fabricate photonic devices, for example lasers with quantum dot active layers which exhibit increased temperature stability, high efficiency, and low threshold current [2, 3]. Quantum dots could also lead to novel device applications in optoelectronics or quantum cryptography. For a comprehensive review, see [4].

1.2 Self-Organized Fabrication

Technological applications require dense arrays of monodisperse nanoscale structures of identical shape. Initially, the most widely followed approach to forming quantum dots was through electron beam lithography of suitably small-featured patterns. However, besides the fact that the spatial resolution of typically 30 nm is often not small enough, there is still a problem with the large amount of damage introduced during the fabrication process itself. Therefore, self-organized growth mechanisms such as the Stranski–Krastanow growth mode [5] have attracted much interest. Here, a heteroepitaxial film can release its elastic energy by forming small dislocation-free three-dimensional islands [6, 7, 8]. Numerous theoretical studies have investigated the equilibrium shape of these islands [9, 10, 11, 12, 13, 14]. However, it turns out that the evolution of the islands is also strongly influenced by kinetic limitations during growth, i.e. the final result depends strongly on the growth conditions. Better understanding and control of these processes thus require structural characterization.

From above statements, it becomes clear that the mesoscopic length scale plays an outstanding role. First, it is most important with respect to the quantization phenomena mentioned above, which appear below typical length scales of about 20 nm. Secondly, the driving forces during the epitaxial growth of these structures may lead to self-organization processes at these length scales. Thirdly, the long-range strain field that is associated with a mesoscopic structure decays on a mesoscopic length scale. This strain field is believed to be the most important driving force for vertical and lateral spatial ordering of an array of mesoscopic structures.

1.3 X-Ray Diffuse Scattering

The aim of the present book is to highlight the capabilities of X-ray diffuse scattering techniques for the nondestructive characterization of mesoscopic structures. The main advantage of X-ray techniques is based on the high resolution in reciprocal space, which is a direct consequence of the high angular resolution of just a few seconds of arc. Therefore, X-rays are sensitive to very small changes of lattice parameters. Moreover, owing to the coherence lengths of typically a few micrometers, information on electron density fluctuations on a larger length scale than atomic spacings can be obtained. In principle, the entire mesoscopic range can be covered by X-ray scattering.

X-rays interact only weakly with matter. Owing to the small volume fraction of the mesoscopic structures compared with the X-ray penetration depth, the scattering signal is very small and has to be enhanced relative to the large signal from the underlying substrate. Therefore, interface- and surface-sensitive techniques using a grazing-incidence geometry have to be applied. Moreover, highly brilliant synchrotron radiation has to be used.

Since the spot sizes of X-ray beams are typically in the range of a few square millimeters, X-rays average statistically over rather large ensembles of mesoscopic structures. This means that the data are very reliable concerning statistics. However, structural data can be evaluated from experimental results only if a single type of mesoscopic structure predominates. Coexistence of different types of mesoscopic structures – for example quantum dots with different shapes or sizes – seriously complicates a detailed analysis. Since the scattering signal of a single kind of structure is usually widely spread in reciprocal space, the coexistence of different sorts of structures leads to troublesome overlapping of the scattering features. In this sense X-ray techniques are complementary to imaging techniques such as high-resolution transmission electron microscopy [15] and scanning probe microscopy [16]. Those latter methods probe the structure locally with high spatial resolution but often with insufficient statistical significance. Also, analytical electron microscopy techniques [17] such as energy-filtered transmission electron microscopy [18], electron energy loss spectroscopy [19, 20], and energy-dispersive X-ray spectroscopy [21] are very useful. In contrast, real-space-imaging X-ray methods are inadequate for the investigation of mesoscopic structures since the spatial resolution of X-ray tomography [22] or X-ray topography [23] is not better than about one micrometer, and is usually far worse. Although sophisticated techniques have been invented to produce fine X-ray beams [24, 25, 26] that may be scanned across the sample, the spatial resolution is still restricted to values of several hundreds of nanometers.

As long as only small scattering volumes are involved, X-ray diffraction can be described by the quite simple kinematical diffraction theory. This circumstance favors X-ray scattering compared with electron diffraction. The strong interaction of electrons with matter leads to multiple-scattering effects and thus requires treatment by the more complicated dynamical theory. We do not consider here anomalous scattering [27], where the X-ray energy is chosen close to an absorption edge of the sample. If this case is avoided, the amplitude of the scattered wave is proportional to the Fourier transform of the electron density. The exactly periodic part of the electron density gives rise to sharp Bragg peaks, whereas all nonperiodic deviations from the ideal crystal structure lead to diffuse scattering in the vicinity of the Bragg peaks. However, when scattering geometries with grazing-incidence conditions are utilized, the kinematical theory is not valid anymore and needs some modifications. These modifications can be made in the framework of the distorted-wave Born approximation.

Independent of the actual scattering theory, the diffuse intensity contains information on the size, shape, positional correlation, local chemical composition, and strain field of the mesoscopic structures. Such a mixture is not easy to demerge and analyze in detail. It is therefore necessary to combine different X-ray scattering techniques of varying sensitivity to these structural properties. Moreover, owing to the loss of phase information in reciprocal space, the scattered intensities cannot be directly "translated" to complete information in real space. X-ray diffuse-scattering simulations that are based on a specific structural model in real space have to be performed and compared with the experimental data. In sequential steps, the model is then refined until satisfactory agreement is achieved.

1.4 How Is this Book Organized?

Along with the fast developments in growth techniques, the corresponding field of X-ray diffuse scattering has been developed – both experimentally and theoretically – on a short timescale, and excellent research has been undertaken (e.g. [28, 29, 30, 31, 32, 33, 34, 35, 36, 37, 38, 39, 40]). Without claiming completeness, this book tries to give an overview of the present state of the art of X-ray diffuse scattering from mesoscopic structures.

First, an overview of different theories of X-ray diffuse scattering is given. This includes a discussion of a novel kinematical-scattering approach where the finite-element method is utilized to calculate the strain field inside mesoscopic structures. The result is then used as an input for subsequent X-ray scattering simulations. After comparison with the experiment, the structural model – which may include, for example, the shape, size, or composition – can be further refined until satisfactory agreement is achieved. The discussion of kinematical theory is followed by a presentation of the more sophisticated distorted-wave Born approximation, which is needed to describe surface- and interface-sensitive scattering under grazing-incidence conditions. The theory of diffuse scattering from surface and interface roughness is discussed in a separate chapter. Here, analytical models for the height–height correlation function, for example, are used, and the theoretical results can be fitted to the corresponding experimental data. This procedure is thus different from the approach mentioned above of using a structural model in real space which is used to calculate X-ray diffuse scattering and which is sequentially refined until sufficient agreement with experiment is achieved.

The theoretical section is followed by an experimental part. After the requirements and the actual properties of synchrotron radiation X-ray sources are described, the most relevant experimental techniques and scattering schemes are introduced and compared with each other. Special attention is given to multidetection techniques, which allow a very efficient and elegant method of three-dimensional data acquisition in reciprocal space. This is implemented through the use of area detectors with a large dynamic range. The relevance of such measurements has not yet attracted sufficient attention in the literature.

The main part of the book deals with the discussion of selected examples. Among them is a model system of SiGe islands grown by liquid-phase epitaxy (LPE). These islands are too large to show quantum size effects. However, they are fabricated rather close to thermodynamic equilibrium, and these islands are very regular in shape, size, and composition. Beyond the growth aspect – which will be extensively discussed – LPE-grown SiGe islands also serve as a model system for X-ray diffuse scattering. They are sufficiently large to give satisfactorily large scattering signals. On the other hand, the very regular shape and size can be exploited to determine small composition changes inside the islands. LPE-grown SiGe islands are also preeminently suited to visualizing dynamical effects in grazing-incidence scattering/diffraction, and they thus represent ideal candidates to test the distorted-wave Born approximation.

Another important emphasis of this book is the investigation of the diffuse scattering from quantum dot ensembles. We try to demonstrate the capability of X-ray diffuse scattering to obtain information about the shape, size, and lateral and vertical positional correlation of the quantum dots, as well as about the chemical composition. For example, InP/InGaP quantum dots show a distinct shape asymmetry. X-ray diffuse scattering along with corresponding simulations can determine the three-dimensional aspect ratios of the base widths and height. These quantities are often used as an input for the calculation of exciton binding energies.

Finally, interfacial roughness is discussed. The existence of atomic steps and step bunches can generate anisotropic interface morphologies. It can also induce inclined inheritance of interface roughness in a multilayer. It will be shown that with increased strain, the model of self-affine roughness can break down, and well-ordered regular, monodisperse step bunches show up that may partially release their internal strain. We shall conduct a critical study of how these effects can be analyzed by X-ray diffuse scattering.

References

1. P. M. Petroff, A. Lorke, and A. Imamoglou, Phys. Today **5**, 46 (2001).
2. Y. Arakawa and H. Sakaki, Appl. Phys. Lett. **40**, 939 (1982).
3. F. Klopf, J. P. Reithmayer, and A. Forchel, Appl. Phys. Lett. **77**, 1419 (2000).
4. D. Bimberg, M. Grundmann, and N. N. Ledentsov, *Quantum Dot Heterostructures* (Wiley, Chichester, 1999).
5. I. Stranski and L. Krastanow, Sitzungsberichte d. Akademie d. Wissenschaften in Wien, Abt. IIb **146**, 797 (1937).
6. D. J. Eaglesham and M. Cerullo, Phys. Rev. Lett. **64**, 1943 (1990).
7. Y.-W. Mo, D. E. Savage, B. S. Swartzentruber, and M. G. Lagally, Phys. Rev. Lett. **65**, 1020 (1990).
8. J. Tersoff and F. K. LeGoues, Phys. Rev. Lett. **72**, 3570 (1994).
9. N. Moll, M. Scheffler, and E. Pehlke, Phys. Rev. B **58**, 4566 (1998).
10. E. Pehlke, N. Moll, A. Kley, and M. Scheffler, Appl. Phys. A **65**, 525 (1997).
11. L. G. Wang, P. Kratzer, N. Moll, and M. Scheffler, Phys. Rev. B **62**, 1897 (2000).

12. L. G. Wang, P. Kratzer, M. Scheffler, and N. Moll, Phys. Rev. Lett. **82**, 4042 (1999).
13. Q. K. K. Liu, N. Moll, M. Scheffler, and E. Pehlke, Phys. Rev. B **60**, 17008 (1999).
14. J. Tersoff and R. M. Tromp, Phys. Rev. Lett. **70**, 2782 (1993).
15. D. B. Williams and C. B. Carter, *Transmission Electron Microscopy – A Textbook for Materials Science* (Plenum, New York, 1996).
16. H. Eisele, O. Flebbe, T. Kalka, C. Preinesberger, F. Heinrichsdorff, A. Krost, D. Bimberg, and M. Dähne-Prietsch, Appl. Phys. Lett. **75**, 106 (1999).
17. J. J. Hren and J. I. Goldstein, *Introduction to Analytical Electron Microscopy* (Plenum, New York, 1979).
18. L. Reimer, *Energy-Filtering Transmission Electron Microscopy* (Springer, Berlin, Heidelberg, 1995).
19. R. F. Egerton, *Electron Energy-Loss Spectroscopy in the Electron Microscope* (Plenum, New York, 1996).
20. R. Schneider, *Electron Energy Loss Spectroscopy (EELS)* (Wiley-VCH, Weinheim, 2002), pp. 50–70, Surface and Thin Film Analysis.
21. R. Schneider, *Energy-dispersive X-Ray Spectroscopy (EDXS)* (Wiley-VCH, Weinheim, 2002), pp. 194–208, Surface and Thin Film Analysis.
22. P. Cloetens, W. Ludwig, J. Baruchel, D. van Dyck, J. van Landuyt, J. P. Guigay, and M. Schlenker, Appl. Phys. Lett. **75**, 2912 (1999).
23. R. Köhler and P. Schäfer, Cryst. Res. Technol. **37**, 734 (2002).
24. A. Snigirev, Rev. Sci. Instrum. **66**, 2053 (1995).
25. S. D. Fonzo, W. Jark, S. Lagomarsino, C. Giannini, L. D. Caro, A. Cedola, and M. Müller, Nature **403**, 638 (2000).
26. D. Bilderback, S. Hoffman, and D. Thiel, Science **263**, 201 (1994).
27. Y. Waseda, *Anomalous X-Ray Scattering for Materials Characterization*, vol. 179 of *Springer Tracts in Modern Physics* (Springer, Berlin, Heidelberg, 2002).
28. G. H. Vineyard, Phys. Rev. B **26**, 4146 (1982).
29. S. K. Sinha, E. B. Sirota, S. Garoff, and H. B. Stanley, Phys. Rev. B **38**, 2297 (1988).
30. T. Salditt, T. Metzger, and J. Peisl, Phys. Rev. Lett. **73**, 2228 (1994).
31. A. J. Steinfort, P. M. L. O. Scholte, A. Ettema, F. Tuinstra, M. Nielsen, E. Landemark, D.-M. Smilgies, R. Feidenhans'l, G. Falkenberg, L. Seehofer, et al., Phys. Rev. Lett. **77**, 2009 (1996).
32. G. Springholz, V. Holý, M. Pinczolits, and G. Bauer, Science **282**, 734 (1998).
33. A. A. Darhuber, J. Zhu, V. Holý, J. Stangl, P. Mikulík, K. Brunner, G. Abstreiter, and G. Bauer, Appl. Phys. Lett. **73**, 1535 (1998).
34. M. Rauscher, R. Paniago, H. Metzger, Z. Kovats, J. Domke, J. Peisl, H.-D. Pfannes, J. Schulze, and I. Eisele, J. Appl. Phys. **86**, 6763 (1999).
35. V. Holý, U. Pietsch, and T. Baumbach, *High Resolution X-Ray Scattering from Thin Films and Multilayers*, vol. 149 of *Springer Tracts in Modern Physics* (Springer, Berlin, Heidelberg, 1999).
36. G. T. Baumbach, D. Lübbert, and M. Gailhanou, J. Phys. D **32**, A208 (1999).
37. I. Kegel, T. H. Metzger, A. Lorke, J. Peisl, J. Stangl, G. Bauer, J. M. Garcia, and P. M. Petroff, Phys. Rev. Lett. **85**, 1694 (2000).
38. J. Grenzer, N. Darowski, U. Pietsch, A. Daniel, S. Rennon, J. P. Reithmaier, and A. Forchel, Appl. Phys. Lett. **77**, 4277 (2000).
39. J. Stangl, A. Daniel, V. Holý, T. Roch, G. Bauer, I. Kegel, T. H. Metzger, T. Wiebach, T. Schmidt, and O. G. Eberl, Appl. Phys. Lett. **79**, 1474 (2001).
40. R. Lazzari, J. Appl. Crystallogr. **35**, 406 (2002).

2 Basic Principles of X-Ray Diffuse Scattering from Mesoscopic Structures

In this chapter we review the most relevant aspects of the theory of X-ray diffuse scattering from mesoscopic structures. Owing to the extensive work that has been undertaken in past years, this cannot be a thorough review but has to be incomplete to some extent. Starting from basic considerations, various theoretical approaches to X-ray scattering, including kinematical theory, dynamical theory, and the distorted-wave Born approximation, are compared. We discuss how X-ray diffuse scattering can be used to extract essential morphological properties of mesoscopic structures, such as shape, size, strain, chemical composition, and positional correlation. Special attention is directed to a novel numerical brute-force technique where elasticity theory is utilized to calculate the strain field inside mesoscopic structures. The resulting strain field is then used as input for subsequent X-ray-scattering simulations. This method allows a systematic refinement of, for example, the shape, size, or composition until satisfactory agreement with experiment is achieved.

Special attention is given to the theory of diffuse scattering from surface and interface roughness, which is described in a separate chapter. In order to describe the surface or interface morphology, sophisticated height–height correlation functions can be employed and the theoretical calculations can be fitted to the corresponding experimental data. In this sense, this procedure is different from the above-mentioned approach of using a structural model in real space which is used to calculate X-ray diffuse scattering and which is – after comparison with experiment – sequentially refined until sufficient agreement is achieved.

2.1 Choice of Appropriate Scattering Theory

The splitting of X-ray diffraction theory into the *dynamical* and *kinematical* approaches has led to the development of two corresponding communities of X-ray physicists. The dynamical approach, which was initially developed by Ewald [1] and von Laue [2], represents the exact solution of Maxwell's equations for the electromagnetic wave fields inside and outside the crystal. It comprises multiple scattering, polarization effects, and refraction/reflection at interfaces. Dynamical theory can be applied to highly perfect systems, e.g.

perfect semi-infinite single crystals, where, as a consequence of the atomic periodicity, narrow Bragg peaks evolve.

Unfortunately, the exact theoretical treatment of X-ray diffuse scattering from mesoscopic structures in the framework of dynamical theory is practically impossible. The mesoscopic structures discussed here show strong deviations from perfect periodicity. Among these deviations are interface roughness, the finite three-dimensional size of a quantum dot, the associated inhomogeneous strain field, and positional correlation between different quantum dots. All these deviations from the ideal system described above give rise to diffuse scattering in the vicinity of the Bragg peaks.

Initially, owing to the lack of perfect crystals, the dynamical theory did not work well, and the experimental results could be better described by the much simpler kinematical theory [3, 4, 5, 6]. The kinematical approach has been successfully applied to describe, for example, diffraction from polycrystalline materials, powder diffraction, and diffraction from soft matter. As a general rule, it is valid for rather imperfect systems and small scattering volumes.

For systems with increasing perfection, the kinematical scattering theory approaches its limitations. This holds especially for the rather perfect semiconductor samples under consideration here. In particular, refraction and multiple scattering are not taken into account. Refraction is usually small at large angles with respect to the surface. It does, however, play a significant role under grazing-incidence and/or grazing-exit conditions. On the other hand, multiple scattering is especially strong for rather perfect regions inside the crystal. These may comprise the perfect substrate or uniformly strained regions inside the barrier layers.

As mentioned above, the structures considered here are too distorted to treat the scattering in the framework of dynamical theory. On the other hand, they are too perfect as to be treated in conventional kinematical scattering theory. How can this dilemma be resolved and how can the diffuse scattering from mesoscopic structures be treated theoretically? One of the main problems is how the diffuse scattering from strongly disturbed regions inside and in the immediate vicinity of the mesoscopic structure is extracted from the scattered signal from the entire sample, which is usually rather perfect. An elegant procedure of separating the entire system into a perfect reference crystal and a distorted part can be performed in the framework of an advanced kinematical theory. Within this approach, the scattering from an undisturbed sample is assumed to be dynamical, while the strongly distorted regions inside the sample can be treated kinematically. This procedure will be described in detail in Sect. 2.3.

An excellent way of taking into account refraction and specular reflection is to calculate the diffuse scattering in first-order perturbation theory, which is also known as the distorted-wave Born approximation (DWBA) [7, 8, 9, 10, 11, 12]. Here, the electron density of our real crystal is separated into a perfect part (reference crystal) and a small perturbation which is responsible for the diffuse scattering. The ideal reference crystal is chosen such that X-ray diffraction from it can be treated within the complete (dynamical) theory. The diffuse scattering is then calculated from corresponding matrix elements that contain both the exact dynamical solutions for the reference crystal and the small perturbation. First, a short outline of dynamical theory is given in Sect. 2.4, which is followed by a detailed description of the DWBA in Sect. 2.5. For an extended review, see the excellent textbook by Holý, Pietsch, and Baumbach [13].

The DWBA is often applied to calculate the diffuse scattering from surface and interface roughness. However, there are specific differences compared with scattering from nanoscale islands. Therefore, diffuse scattering from roughness will be discussed separately in Sect. 2.6.

Before we go into more detail about these theories that are used to describe X-ray diffuse scattering, we start with basic considerations about the interaction of X-rays with matter. These include reflection, refraction, and absorption of X-rays. (Sect. 2.2). Since these basic principles can be found in various excellent textbooks (e.g. [3, 13, 14, 15, 16]) or review articles [17, 18, 19], only the most relevant aspects are reviewed here.

2.2 Interaction of Hard X-Rays with Matter

2.2.1 The Dielectric Polarizability

The interaction of electromagnetic waves with matter is mediated by the complex dielectric[1] polarizability $\chi(\boldsymbol{r},\omega)$. It depends on the X-ray energy $\hbar\omega$ and can be expressed as

$$\chi(\boldsymbol{r},\omega) = -\frac{r_0 N_\mathrm{A} \lambda^2}{\pi A} \varrho_\mathrm{m}(\boldsymbol{r})[f_0 + f'(\omega) - \mathrm{i}f''(\omega)] \,. \tag{2.1}$$

Here $\varrho_\mathrm{m}(\boldsymbol{r})$ is the mass density, N_A is Avogadro's number, A is the molar atomic mass, and λ is the electromagnetic wavelength. f_0 is the atomic form factor (or form amplitude) and is defined as

$$f_0 = \int \varrho_\mathrm{A}(\boldsymbol{r}) \mathrm{e}^{\mathrm{i}\boldsymbol{q}\cdot\boldsymbol{r}} \, \mathrm{d}V \,, \tag{2.2}$$

where $\varrho_\mathrm{A}(\boldsymbol{r})$ is the electron density of a single atom. The energy-dependent factors $f'(\omega)$ and $f''(\omega)$ – also known as Hönl corrections [20] – correct for dispersion and absorption, respectively. They become important at absorption edges and are interconnected by a Kramers–Kronig relation [21, 22, 23]. Typical values of $\mathrm{Re}(\chi)$ and $\mathrm{Im}(\chi)$ for hard X-rays are in the range of 10^{-5} and 10^{-7}, respectively (for solids and liquids).

[1] In this book, we do not deal with (resonant) magnetic scattering or magnetic dichroism, which can be described by the complex magnetic polarizability.

The classical electron radius r_0 – often referred to as the Thomson scattering length – is given by

$$r_0 = \frac{e^2}{4\pi\epsilon_0 m c_0^2} = 2.818 \times 10^{-5}\,\text{Å} \,. \tag{2.3}$$

Equation (2.1) is valid for a material consisting of a single atomic species, i.e. a chemical element. In the case of different atomic species k in the medium (e.g. GaAs or SiGe), the partial mass densities $\varrho_{\mathrm{m},k}$ with corresponding $f_{0,k}$, f'_k, f''_k and partial molecular atomic masses A_k have to be considered, leading to

$$\chi(\boldsymbol{r},\omega) = -\frac{r_0 N_{\mathrm{A}} \lambda^2}{\pi} \sum_k \frac{\varrho_{\mathrm{m},k}(\boldsymbol{r})}{A_k} [f_{0,k} + f'_k(\omega) - \mathrm{i} f''_k(\omega)] \,. \tag{2.4}$$

If we neglect dispersion and absorption edges – thus disregarding resonant anomalous scattering (e.g. [24, 25]) – we can write

$$\chi(\boldsymbol{r},\omega) = \chi(\boldsymbol{r}) = -r_0 \frac{\lambda^2}{\pi} \varrho(\boldsymbol{r}) \,, \tag{2.5}$$

where $\varrho(\boldsymbol{r})$ is the electron density. Since $\chi(\boldsymbol{r})$ follows the periodicity of the crystal, it can be expressed in the form of a Fourier series over the reciprocal lattice:

$$\chi(\boldsymbol{r}) = \sum_{\boldsymbol{h}} \chi_{\boldsymbol{h}} \mathrm{e}^{\mathrm{i}\boldsymbol{h}\cdot\boldsymbol{r}} \,. \tag{2.6}$$

We can then split the dielectric polarizability (and thus also the electron density) into a constant part χ_0 (zeroth Fourier coefficient) and the local deviation $\Delta\chi(\boldsymbol{r})$ from it:

$$\chi(\boldsymbol{r}) = \chi_0 + \Delta\chi(\boldsymbol{r}) \,. \tag{2.7}$$

The latter takes into account the atomic (crystalline) structure, whereas χ_0 represents the average value over the elementary unit cell of the crystal. We shall show later that χ_0 can be used to describe scattering at small values of $\boldsymbol{q}$, whereas the oscillatory part $\Delta\chi(\boldsymbol{r})$ must be used to treat large-angle X-ray diffraction in the vicinity of a nonzero reciprocal-lattice point.

2.2.2 Reflection, Refraction, and Absorption of X-Rays

At small scattering angles the X-rays average over the elementary unit cell of the crystal, i.e. only the constant part χ_0 of the polarizability $\chi(\boldsymbol{r})$ has to be considered. According to (2.1), the complex refractive index n_0 can be expressed as

$$n_0 = (1+\chi_0)^{1/2} \approx 1 + \frac{1}{2}\chi_0 = 1 - \delta + \mathrm{i}\beta \,, \tag{2.8}$$

where

$$\delta = \lambda^2 \frac{r_0 N_A \varrho_m}{2\pi A} (f_0 + f') \tag{2.9}$$

and

$$\beta = \lambda^2 \frac{r_0 N_A \varrho_m}{2\pi A} f'' = \frac{\lambda\mu}{4\pi} \,. \tag{2.10}$$

Here μ denotes the linear absorption coefficient.

Let us now consider an incoming plane wave $\boldsymbol{E}(\boldsymbol{r}, t)$ which illuminates, at a glancing angle α_i (see Fig. 2.1), a flat surface of a medium with a refractive index n_0. The incoming wave is partly reflected at the surface ($\boldsymbol{E}_r(\boldsymbol{r}, t)$) and is partly transmitted into the medium ($\boldsymbol{E}_t(\boldsymbol{r}, t)$). We can write

$$\boldsymbol{E}(\boldsymbol{r}, t) = \boldsymbol{E}_0 e^{i(\boldsymbol{k}\cdot\boldsymbol{r} - \omega t)} \,, \tag{2.11a}$$

$$\boldsymbol{E}_t(\boldsymbol{r}, t) = \boldsymbol{E}_t e^{i(\boldsymbol{k}'\cdot\boldsymbol{r} - \omega t)} \,, \tag{2.11b}$$

$$\boldsymbol{E}_r(\boldsymbol{r}, t) = \boldsymbol{E}_r e^{i(\boldsymbol{k}''\cdot\boldsymbol{r} - \omega t)} \,, \tag{2.11c}$$

where the wave vectors $\boldsymbol{k}$, $\boldsymbol{k}'$, $\boldsymbol{k}''$, the frequency ω, and the refractive index n_0 are interconnected via

$$|\boldsymbol{k}| = |\boldsymbol{k}''| = k = \frac{2\pi}{\lambda} = \frac{\omega}{c_0} \,; \quad |\boldsymbol{k}'| = k' = \frac{2\pi}{\lambda'} = \frac{\omega}{c_0} n_0 \,. \tag{2.12}$$

Here c_0 is the vacuum velocity of light.

The deviation of the refractive index from unity is very small for X-rays. Therefore, significant reflection at surfaces and interfaces appears at very small angles only. However, since the refractive index is smaller than unity, total external reflection occurs below the X-ray critical angle of total external

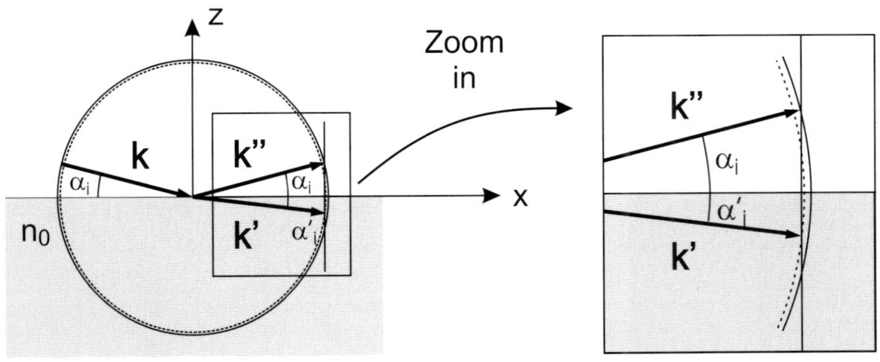

Fig. 2.1. Reflection and refraction of an X-ray plane wave at a surface. The radii of the Ewald spheres for the incoming/reflected and transmitted waves (sketched as *solid* and *dashed lines*, respectively) are given by $2\pi/\lambda$ and $2\pi n_0/\lambda$, respectively, λ being the X-ray vacuum wavelength. Since the horizontal components of the wave vectors $\boldsymbol{k}$, $\boldsymbol{k}'$, and $\boldsymbol{k}''$ are preserved, the transmitted wave is refracted

reflection[2], which is given by

$$\alpha_c = (2\delta)^{1/2} , \tag{2.13}$$

and takes values in the range of a fraction of a degree (see Table 2.1). For angles of incidence below the critical angle, i.e. $\alpha_i \le \alpha_c$, the transmitted wave has a purely imaginary wave vector $\boldsymbol{k} = -i\kappa \boldsymbol{e}_z$, and an exponentially damped evanescent wave field is generated below the surface with a penetration depth Λ in the order of 10–100 Å. As a consequence total external reflection occurs. On the other hand, for values $\alpha_i > \alpha_c$, the transmitted wave may enter the medium and is attenuated only by absorption, which can be described by the linear absorption coefficient as defined in (2.10). The penetration depth Λ can be generally expressed as [29]

$$\Lambda = \frac{\lambda}{2\pi l} , \tag{2.14}$$

where

$$l = 2^{-1/2}\{(2\delta - \sin^2\alpha_i) + [(\sin^2\alpha_i - 2\delta)^2 + 4\beta^2]^{1/2}\}^{1/2} . \tag{2.15}$$

Table 2.1. Refractive index $n_0 = 1 - \delta + i\beta$ and critical angle α_c for different materials at $\lambda = 1.54$ Å. The data have been taken from [30][a]

Material	δ (10^{-5})	β (10^{-7})	β/δ	α_c
Si	0.7563	1.748	0.023	0.22°
GaAs	1.441	4.087	0.028	0.31°
Ge	1.444	4.200	0.029	0.31°

[a] S.A. Stepanov has written the program X0h, which is available at [30]. The databases and interpolation procedures are also available at this website.

For small values of α_i and α_c, the sine function can be replaced by its argument and, as a consequence, $\alpha_c\Lambda$ scales with β/δ. The corresponding plot is shown in Fig. 2.2. At $\lambda = 0.154$ nm the minimum penetration depths for Si and GaAs are $\Lambda_{Si} = 6.4$ nm and $\Lambda_{GaAs} = 4.5$ nm, respectively.

The amplitudes of the reflected and transmitted waves can be calculated by using the Fresnel equations. For small angles of incidence compared with the Brewster angle, the Fresnel transmission and reflection coefficients are independent of the polarization state of the X-rays [31] and can be written as

$$t_F := \frac{E_t}{E_0} = \frac{2k_z}{k_z + k'_z} , \qquad r_F := \frac{E_r}{E_0} = \frac{k_z - k'_z}{k_z + k'_z} = 1 - t_F . \tag{2.16}$$

[2] As early as 1922, A.H. Compton suggested to use this relationship to measure the refractive index of X-rays [26, 27, 28].

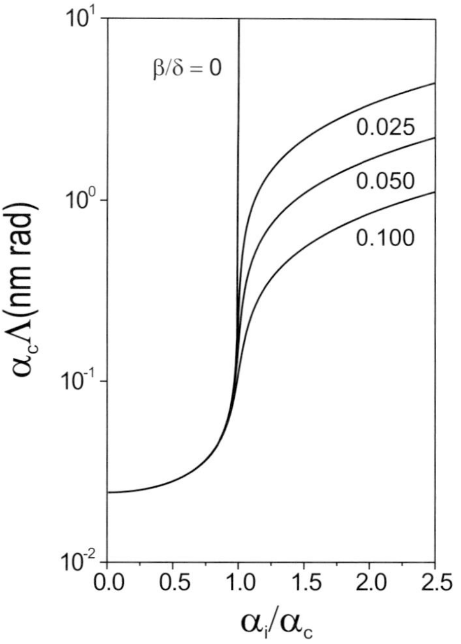

Fig. 2.2. $\alpha_c \Lambda$ as a function of α_i/α_c for different values of δ and β. For Si, Ge, and GaAs, typical values are $\beta/\delta = 0.025$ at $\lambda = 1.54$ Å (see also Table 2.1)

Here k_z and k'_z are the z-components of the wave vectors outside and inside the crystal, respectively. The Fresnel coefficients r_F and t_F can be also expressed in terms of the glancing angles α_i and α'_i of the incoming and refracted waves, respectively, with respect to the sample surface (or to an interface between two layers), through

$$r_F = \frac{\sin\alpha_i - n_0 \sin\alpha'_i}{\sin\alpha_i + n_0 \sin\alpha'_i}\,, \qquad t_F = \frac{2\sin\alpha_i}{\sin\alpha_i + n_0 \sin\alpha'_i}\,. \tag{2.17}$$

Note that r_F and t_F are defined with respect to the amplitudes E_0, E_t, and E_r of the incoming, transmitted, and reflected plane waves, respectively. Generally, they are complex numbers, since the refractive index n_0 is complex. The experimentally accessible quantities are, however, not the amplitudes but the reflected and transmitted intensities $|r_F|^2$ and $|t_F|^2$, respectively (Fig. 2.3).

Equation (2.16) describes the reflectivity of a single, flat surface. When one is treating the reflectivity of an arbitrary multilayer, the amplitudes of the waves that are reflected and transmitted at all interfaces have to be coherently summed. Parratt [32] has deduced a recurrence formula that iteratively relates the reflected and transmitted amplitudes, R_j and T_j, respectively, of

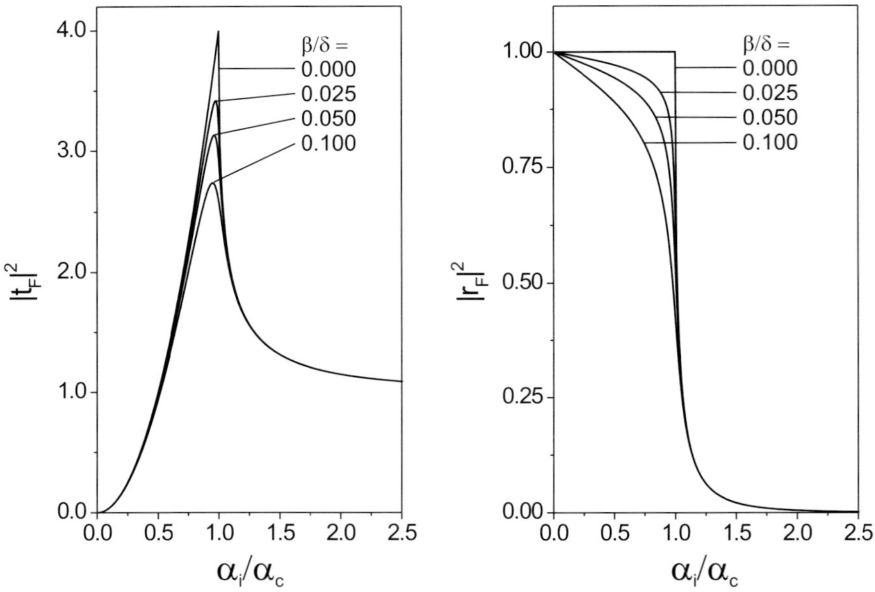

Fig. 2.3. Fresnel transmittivity $|t_{\mathrm{F}}|^2$ and reflectivity $|r_{\mathrm{F}}|^2$ as a function of $\alpha_{\mathrm{i}}/\alpha_{\mathrm{c}}$

a multilayer (Fig. 2.4) with interfaces labeled by j at positions z_j via [33]

$$X_j = \frac{R_j}{T_j} = \mathrm{e}^{-2\mathrm{i}k_{z,j}z_j} \frac{r_{j,j+1} + X_{j+1}\mathrm{e}^{2\mathrm{i}k_{z,j}z_j}}{1 + r_{j,j+1}X_{j+1}\mathrm{e}^{2\mathrm{i}k_{z,j}z_j}} , \tag{2.18}$$

where the Fresnel reflection coefficient of interface j is given by

$$r_{j,j+1} = \frac{k_{z,j} - k_{z,j+1}}{k_{z,j} + k_{z,j+1}} . \tag{2.19}$$

This formula can be iteratively solved if we make use of the fact that there is no reflection from the lower interface of the substrate ($R_{N+1} = 0$) and by normalizing the reflectivity with respect to the incoming wave ($T_1 = 1$). It is noteworthy that (2.19) takes into account multiple reflection at all interfaces, although it is only important at angles close to α_{c}.

An equivalent description has been developed by Abelès [34, 35], who introduced transfer matrices that relate the amplitudes of the reflected and transmitted waves from layers j and $j+1$ (Fig. 2.4):

$$\begin{pmatrix} A_j \\ A_j'' \end{pmatrix} = \mathsf{M}^{(j,j+1)} \begin{pmatrix} A_{j+1} \\ A_{j+1}'' \end{pmatrix} . \tag{2.20}$$

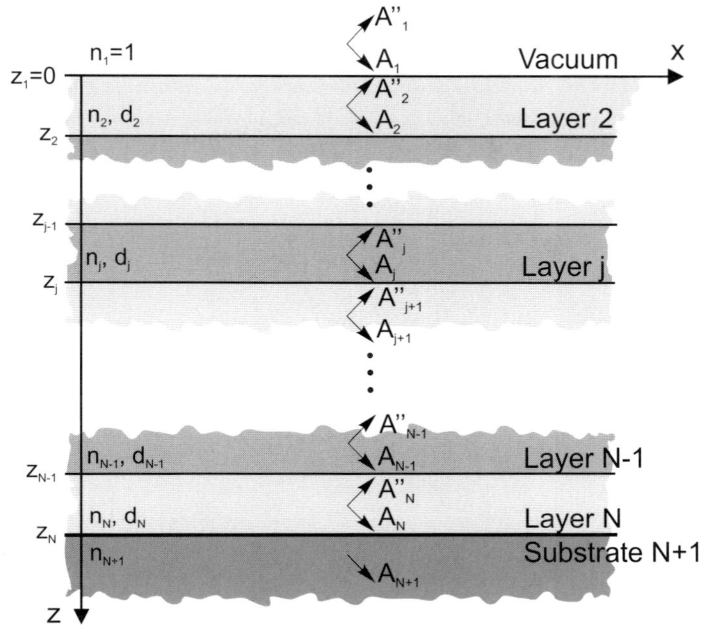

Fig. 2.4. Illustration of a multilayer with ideally flat interfaces. The multilayer consists of N layers (and N interfaces) that are grown on a substrate. Each layer is homogeneous and is characterized by its refractive index n_j and thickness d_j. The interface j is the lower limit of the layer j. $z = 0$ corresponds to the surface. A''_j is the reflected amplitude from interface j, and A_j is the transmitted amplitude inside layer j. The figure has been taken from Opitz [36]

The matrix elements of M are given by [37]

$$M_{11}^{(j,j+1)} = \frac{\theta_j + \theta_{j+1}}{2\theta_j} \exp\left(-\frac{\mathrm{i}}{2}k(\theta_j d_j + \theta_{j+1} d_{j+1})\right) , \tag{2.21a}$$

$$M_{12}^{(j,j+1)} = \frac{\theta_j - \theta_{j+1}}{2\theta_j} \exp\left(-\frac{\mathrm{i}}{2}k(\theta_j d_j - \theta_{j+1} d_{j+1})\right) , \tag{2.21b}$$

$$M_{21}^{(j,j+1)} = \frac{\theta_j - \theta_{j+1}}{2\theta_j} \exp\left(\frac{\mathrm{i}}{2}k(\theta_j d_j - \theta_{j+1} d_{j+1})\right) , \tag{2.21c}$$

$$M_{22}^{(j,j+1)} = \frac{\theta_j + \theta_{j+1}}{2\theta_j} \exp\left(\frac{\mathrm{i}}{2}k(\theta_j d_j + \theta_{j+1} d_{j+1})\right) , \tag{2.21d}$$

where θ_j is the (complex) glancing angle of the jth wave with respect to the jth interface and can be calculated by using Snell's law:

$$\theta_j = \sqrt{\theta_1^2 - 2\delta_j + 2\mathrm{i}\beta_j} \, . \tag{2.22}$$

Note that (2.21) and (2.22) are given in the small-angle approximation, i.e. $\sin(x)$ is replaced by its argument x, and the absolute values of the wave vectors k_j'' inside the multilayer are replaced by the vacuum value k.

The advantage of the transfer matrix formalism can be seen when we consider the totally reflected and transmitted amplitudes A_1'' and A_1, respectively, which can be now easily calculated from

$$\begin{pmatrix} A_1 \\ A_1'' \end{pmatrix} = \mathsf{M}^{(1,2)}\mathsf{M}^{(2,3)}\ldots\mathsf{M}^{(N,N+1)}\begin{pmatrix} A_{N+1} \\ 0 \end{pmatrix} \equiv \mathsf{T}\begin{pmatrix} A_{N+1} \\ 0 \end{pmatrix} . \tag{2.23}$$

Here, we have used the fact that there is no wave coming back from the substrate ($A_{N+1}'' = 0$). The total reflectivity (normalized with respect to the incoming intensity) can be then expressed as

$$R = \left| \frac{A_1''}{A_1} \right|^2 = \left| \frac{T_{21}}{T_{11}} \right|^2 . \tag{2.24}$$

2.2.3 Effect of Roughness on Reflectivity and Transmittivity

So far, ideally flat interfaces have been treated. For interfaces with a finite width or a finite interfacial root mean square (rms) roughness σ, the intensities of the specularly reflected wave and of the transmitted wave are weakened.[3] Therefore, the Fresnel transmission and reflection coefficients t_F and r_F are not valid anymore. There has been a lot of discussion in the literature on how the expressions for t_F and r_F have to be modified. Since only the vertical component of the scattering vector enters into the formula for the reflectivity, the specular beam averages over the lateral structure of roughness. This averaging is mostly treated as being Gaussian.[4]

The crucial point is that roughness with high spatial frequency destroys the coherence between the reflected and transmitted waves. This means that the fixed phase relation between the incoming and reflected waves is lost. In this case the Gaussian average should be calculated separately for the amplitudes of the incoming and reflected (or transmitted) waves. Névot and Croce [41] and others [42, 43] have derived a self-consistent solution for the modified reflection and transmission coefficients,

$$r = r_\mathrm{F}\mathrm{e}^{-2k_z k_z' \sigma^2} \quad \text{and } t = t_\mathrm{F}\mathrm{e}^{(k_z - k_z')^2\sigma^2/2} . \tag{2.25}$$

[3] The cases of interfacial roughness and vertical intermixing (transition layer) at the interfaces lead to the same result and are thus not distinguishable by measuring the reflectivity. As pointed out by Stepanov and Köhler [38], this is also the case for the coherent part of grazing-incidence diffraction. They can, however, be separated by measuring the diffuse scattering.

[4] A Gaussian averaging leads to an effective profile of the refractive index of $n(z) = (1/2)(\mathrm{erf}(z/\sqrt{2}\sigma)+1)$, where $\mathrm{erf}(z) = (2/\sqrt{\pi})\int_0^z \mathrm{d}t\, \exp(-t^2)$ is the error function. This is a sufficiently good approximation. However, other profiles have also been proposed for $n(z)$ [39, 40].

Note that the modified reflectivity and transmittivity are invariant under the exchange of the incoming and transmitted waves, i.e. the reciprocity principle is fulfilled.

Besides this result, the Debye–Waller-like factor

$$r = r_{\mathrm{F}} \mathrm{e}^{-2\sigma^2 k_z^2} \tag{2.26}$$

has often been mentioned in the literature [44, 45], which may give somewhat different results. The Debye–Waller factor was proposed earlier in optics by Rayleigh [46] and is a result of the remaining coherence between the incoming and reflected waves.

De Boer [47] has shown that both cases, the result of Névot and Croce and the Debye–Waller factor, are limiting cases of a more general form that can be obtained by using second-order perturbation theory (see also Sect. 2.5). At low roughness correlation lengths, the result given in (2.25) yields the better approximation, whereas at large correlation lengths the Debye–Waller-like term of (2.26) gives the better description for r. De Boer also proposed a more general expression for the modified transmittivity t. A comprehensive discussion of this problem can be found in [47, 48].

Using the model of Névot and Croce, the recursion formalism of Parratt given in (2.18) can be now generalized, by modifying the Fresnel reflection coefficient in (2.19) by including a term $\mathrm{e}^{-2k_{z,j}k_{z,j+1}\sigma_{j,j+1}^2}$, where $\sigma_{j,j+1}$ is the rms interface roughness of the interface between layers j and $j+1$. Also, the matrix formalism described above can be generalized to rough interfaces. When the model of Névot and Croce is used, the amplitudes have to be averaged separately, which is equivalent to a procedure in which the elements of the transfer matrices have to be averaged. Klemradt [49] has given the result as

$$\langle M_{mn}^{j,j+1} \rangle = M_{mn}^{j,j+1} \exp\left[-\frac{1}{2} k^2 (\theta_j - (-1)^{m+n} \theta_{j+1})^2 \sigma_{j,j+1}^2 \right] . \tag{2.27}$$

2.3 Kinematical Scattering Theory

2.3.1 The Born Approximation

Several mechanisms for the interaction of X-rays with matter are possible, e.g. magnetic interaction or inelastic Compton scattering. Here, only elastic charge scattering (Thomson scattering [50]) will be considered. From a classical point of view, a charged free electron is accelerated by the incoming electromagnetic wave. The resulting forced oscillation of the electron leads to reemission of an electromagnetic spherical wave, which has the same frequency as the original incoming wave and exhibits a characteristic dipole-like angular distribution. At large distance R of the electron from the observer, the electric far field of the scattered wave is given by [13]

$$E_{\rm rad}(\boldsymbol{r}, R) = r_0 P E_0 \mathrm{e}^{\mathrm{i}\boldsymbol{k}^{\mathrm{i}}\cdot\boldsymbol{r}} \frac{\mathrm{e}^{\mathrm{i}kR}}{R} , \qquad (2.28)$$

where $E_0 \mathrm{e}^{\mathrm{i}\boldsymbol{k}^{\mathrm{i}}\cdot\boldsymbol{r}}$ is the incoming wave, r_0 the classical electron radius and P the polarization factor. Since the interaction is elastic, the lengths of the wave vectors are conserved, and we can write $|\boldsymbol{k}^{\mathrm{i}}| = k = 2\pi/\lambda$.

The polarization factor P depends on the scattering geometry and can be generally expressed as

$$P = \begin{cases} 1 & \text{s polarization} \\ |\cos(2\theta)| & \text{p polarization} \\ \sqrt{\frac{1}{2}[1+\cos^2(2\theta)]} & \text{unpolarized light} \end{cases} , \qquad (2.29)$$

where 2θ is the angle between the incident and scattered waves (scattering angle). In the s and p polarizations, the incoming electric field vector is perpendicular and parallel, respectively, to the scattering plane. Most of the experiments described in this book were carried out at synchrotron radiation sources, which emit highly polarized light. Usually, the scattering plane is chosen perpendicular to the electric field vector so that $P = 1$. However, in the other scattering geometries, the polarization factor does not substantially vary within the small areas of reciprocal space of interest. Therefore, the scattering-angle dependence of the polarization factor can be neglected.

Equation (2.28) describes the scattering from a single free electron. How can the scattered amplitude from an arbitrary ensemble of bound electrons (such as that in a crystal) now be derived? As long as the frequency of the electromagnetic field is much larger than the characteristic atomic frequencies, the electrons can be considered as free electrons. More precisely, it is sufficient to be far from all absorption edges, thus avoiding anomalous scattering. In this case the Hönl corrections can be omitted and the material can be characterized simply by its electron density $\varrho(\boldsymbol{r})$.

In the kinematical approach, also known as the first Born approximation, multiple scattering is neglected and the total scattered amplitude is given by a coherent summation over all scattered partial waves,

$$E_{\rm rad}(\boldsymbol{R}) = E_0 r_0 P \int \varrho(\boldsymbol{r}) \mathrm{e}^{\mathrm{i}\boldsymbol{k}^{\mathrm{i}}\cdot\boldsymbol{r}} \frac{\mathrm{e}^{\mathrm{i}k|\boldsymbol{r}-\boldsymbol{R}|}}{|\boldsymbol{r}-\boldsymbol{R}|} \mathrm{d}V . \qquad (2.30)$$

The integration is performed over the coherently illuminated volume. We shall now use the Fraunhofer approximation, which is valid when the distance between the sample and detector is much larger than the diameter of the coherently illuminated sample area,[5] i.e. $|\boldsymbol{R}| \gg |\boldsymbol{r}|$. In this case we can make the approximation $k|\boldsymbol{r}-\boldsymbol{R}| \approx kR - \boldsymbol{k}^{\mathrm{f}}\cdot\boldsymbol{r}$, and (2.30) can be written as

[5] The diameter of the first Fresnel zone of the sample is the decisive parameter. Assuming a perfectly collimated beam ($\lambda = 1.5$ Å) and a distance between the

$$A(\boldsymbol{q}) = A_0 r_0 P \frac{\mathrm{e}^{\mathrm{i}kR}}{R} \int \varrho(\boldsymbol{r}) \mathrm{e}^{\mathrm{i}\boldsymbol{q}\cdot\boldsymbol{r}} \, \mathrm{d}V \ . \tag{2.31}$$

Since the electric and magnetic field amplitudes $E(\boldsymbol{r})$ and $B(\boldsymbol{r})$ are equivalent for a vacuum electromagnetic wave, the generalized term $A(\boldsymbol{q})$ is here used for the amplitude. Also, the vector character of $\boldsymbol{E}$ is omitted.

The phase difference between the scattered waves originating from two electrons that are separated by a distance vector $\boldsymbol{r}$ is given by $\boldsymbol{q} \cdot \boldsymbol{r}$, where $\boldsymbol{q}$ is the scattering vector, which is defined as

$$\boldsymbol{q} = \boldsymbol{k}^{\mathrm{f}} - \boldsymbol{k}^{\mathrm{i}} \ . \tag{2.32}$$

As can be seen from (2.31), the result of the Fraunhofer approximation is that the scattering object is treated as a pointlike source that emits a spherical wave front $\mathrm{e}^{\mathrm{i}kR}/R$, which is modulated by $\int \varrho(\boldsymbol{r})\mathrm{e}^{\mathrm{i}\boldsymbol{q}\cdot\boldsymbol{r}} \, \mathrm{d}V$. Therefore, the wave vector $\boldsymbol{k}^{\mathrm{f}}$ of the scattered radiation is determined by the position $\boldsymbol{R}$ of the detector only. The amplitude $\int \varrho(\boldsymbol{r})\mathrm{e}^{\mathrm{i}\boldsymbol{q}\cdot\boldsymbol{r}} \, \mathrm{d}V$ is only nonzero when $\boldsymbol{q} = \boldsymbol{h}$, where $\boldsymbol{h}$ is a reciprocal-lattice vector; this condition is the Bragg condition. The Bragg-diffracted wave can therefore be treated as a plane wave with a well-defined wave vector $\boldsymbol{k}^{\mathrm{f}}$.

Absorption and refraction can be easily implemented in the kinematical approach by defining a complex scattering vector $\boldsymbol{q}$, where the real part is corrected for refraction and the imaginary part describes absorption. For a plane wave illuminating a crystal at normal incidence, the wave vector changes from $k = 2\pi/\lambda$ in vacuum to $k' = n_0 k$ in the crystal. Since the refractive index n_0 is a complex number, i.e. $n_0 = 1 - \delta + \mathrm{i}\beta$, the wave propagating in the $-z$ direction[6] into the crystal can be expressed as $\mathrm{e}^{-\mathrm{i}n_0 kz} = \mathrm{e}^{-\mathrm{i}(1-\delta)kz}\mathrm{e}^{\beta kz} = \mathrm{e}^{-\mathrm{i}(1-\delta)kz}\mathrm{e}^{\mu z/2}$. Since relative changes of scattered intensity are probed in just a small area of reciprocal space, absorption can usually be neglected as long as the photon energy is not chosen close to an absorption edge. Usually, refraction is not treated within the kinematical approach but within the framework of the distorted-wave Born approximation (see Sect. 2.5).

If the prefactors in front of the integral in (2.31) are omitted, $A(\boldsymbol{q})$ is the three-dimensional Fourier transform of the electron density $\varrho(\boldsymbol{r})$. However, in the experiment intensities are measured. These are related to the differential scattering cross section, which describes the normalized scattered photon flux

sample and detector of $D = 1\,\mathrm{m}$, one can easily estimate the diameter of the first Fresnel zone as $d = 2\sqrt{D\lambda/2} \approx 17\,\mu\mathrm{m}$, which is much smaller than the X-ray beam diameter at the sample. However, the area of the Fresnel zone is still larger than the coherently illuminated area on the sample, which is typically some $\mu\mathrm{m}^2$ (see also Sect. 3.3).

[6] Note that in Fig. 2.1, the z-axis points outwards from the surface. Therefore, the term $\mathrm{e}^{\mu z/2}$ is equivalent to exponential *damping* of the amplitude inside the crystal.

into an elementary solid angle $\mathrm{d}\Omega$:

$$\frac{\mathrm{d}\sigma}{\mathrm{d}\Omega}(\boldsymbol{q}) = \frac{|A(\boldsymbol{q})|^2}{|A_0|^2} R^2 = r_0^2 P^2 \left| \int \varrho(\boldsymbol{r}) \mathrm{e}^{\mathrm{i}\boldsymbol{q}\cdot\boldsymbol{r}} \, \mathrm{d}V \right|^2 . \tag{2.33}$$

Using the convolution theorem, this can be expressed as

$$\frac{\mathrm{d}\sigma}{\mathrm{d}\Omega}(\boldsymbol{q}) = r_0^2 P^2 \int C(\boldsymbol{r}) \mathrm{e}^{\mathrm{i}\boldsymbol{q}\cdot\boldsymbol{r}} \, \mathrm{d}V , \tag{2.34}$$

where $C(\boldsymbol{r})$ is the *autocorrelation function*,

$$C(\boldsymbol{r}) = \int \varrho(\boldsymbol{r}')\varrho(\boldsymbol{r}+\boldsymbol{r}') \, \mathrm{d}V' . \tag{2.35}$$

Equation (2.34) reflects one of the most basic features of kinematical diffraction: the total intensity is related to the Fourier transform of an autocorrelation function. This correlation function has to be considered at different length scales, and it may describe both short-range and long-range ordering. For example:

- The autocorrelation function $C(\boldsymbol{r})$ can describe spatial correlation phenomena on the mesoscopic length scale, e.g. positional ordering in an array of quantum dots, which can usually be described by a short-range-order model. A quantitative measure of the state of ordering is given by the correlation length ξ, which will be described later.
- $C(\boldsymbol{r})$ includes short- and long-range ordering phenomena on atomic length scales. Among these phenomena is the strain field inside and in the proximity of mesoscopic structures. Also, the perfect part of the substrate lattice can be described by $C(\boldsymbol{r})$.
- Electron density fluctuations can be characterized by $C(\boldsymbol{r})$. These may be present over the entire mesoscopic length scale. Examples are the vertical electron density profile in a superlattice, and local composition changes inside a quantum dot.
- In (2.34), the shape and size of mesoscopic structures are implicitly included.

In the following sections we shall discuss in detail how strain, shape, size, and positional correlation enter into the diffusely scattered intensities.

2.3.2 Strain-Induced Diffuse Scattering

In a perfect periodic lattice, (2.34) leads to sharp Bragg peaks in reciprocal space. Any perturbation of the exact periodicity of the atomic lattice gives rise to diffuse scattering in the vicinity of these Bragg peaks. In order to evaluate the diffusely scattered intensity, the real crystal with electron density $\varrho_{\mathrm{real}}(\boldsymbol{r})$ is separated into two parts,

$$\varrho_{\text{real}}(\boldsymbol{r}) = \varrho_{\text{dif}}(\boldsymbol{r}) + \varrho_{\text{ref}}(z) \; . \tag{2.36}$$

The second term on the right-hand side of (2.36) describes an ideal, laterally homogeneous crystal, which is called the *reference lattice*. The difference $\varrho_{\text{dif}}(\boldsymbol{r}) = \varrho_{\text{real}}(\boldsymbol{r}) - \varrho_{\text{ref}}(z)$ between the real sample and this reference lattice is treated as a small perturbation with respect to the ideal part.

The majority of the total intensity is well described by diffraction from the perfect reference lattice, which does not produce any diffuse scattering. Scattering from the reference lattice, therefore, has to be treated dynamically. In contrast, the weak diffuse scattering originating from the mesoscopic structures is treated kinematically. In the case of free-standing islands grown on a perfect substrate, the substrate lattice is chosen as the reference lattice (Fig. 2.5).

Since the mesoscopic structures are only rather small fractions of the illuminated crystal volume, the diffuse scattering has to be ascribed to the difference between the real and the reference lattice:

$$A^{\text{diffuse}}(\boldsymbol{q}) = A_0 r_0 P \frac{\mathrm{e}^{\mathrm{i}kR}}{R} \int \varrho_{\text{dif}}(\boldsymbol{r}) \mathrm{e}^{\mathrm{i}\boldsymbol{q}\cdot\boldsymbol{r}} \, \mathrm{d}V \; . \tag{2.37}$$

By taking the proper local amplitudes for the diffraction from the reference lattice ($\varrho_{\text{ref}}(z)$), the diffraction from the mesoscopic structures ($\varrho_{\text{dif}}(\boldsymbol{r})$) can be described by kinematical scattering theory as outlined above.

The real sample needs to be described in terms of the electron density. Since most of the electrons are tightly bound to the ion cores, the local electron density inside an atom is practically insensitive to strain. Consequently,

$$\varrho_{\text{real}}(\boldsymbol{r}) = \varrho_{\text{ideal}}(\boldsymbol{r} - \boldsymbol{u}(\boldsymbol{r})) \tag{2.38}$$

should be a good approximation, where $\boldsymbol{u}(\boldsymbol{r})$ is the displacement field of the atoms in the real crystal compared with the positions of the atoms in an ideal crystal. This approximation is crucial for the validity of the present theory. It is not obvious, since strain modifies the band structure and, consequently, the electron density. Kuriyama [51] found that (2.38) is valid if the deformation is so small that there is no overlap of the core electron wave functions of the displaced atoms. In addition, the number of valence electrons must be a minor fraction of all scattering electrons. For strong reflections this is always fulfilled for the material systems investigated in the work described here.

It is worth mentioning, however, that the situation changes when weak (quasi-forbidden) reflections are investigated. Reflections of the type $h00$ ($h = 2, 6, 10, \ldots$) are weak in the zinc blende structure because the total scattered intensity depends on the difference between the atomic form amplitudes only. Although (2.38) has been successfully applied to AlGaAs/GaAs systems [52, 53, 54, 55], the validity should be questioned when one is treating strongly strained systems, e.g. InAs/GaAs.

Usually, the lattice of the unstrained substrate is chosen as the ideal crystal. Therefore, the ideal crystal does not generally exhibit the same lattice

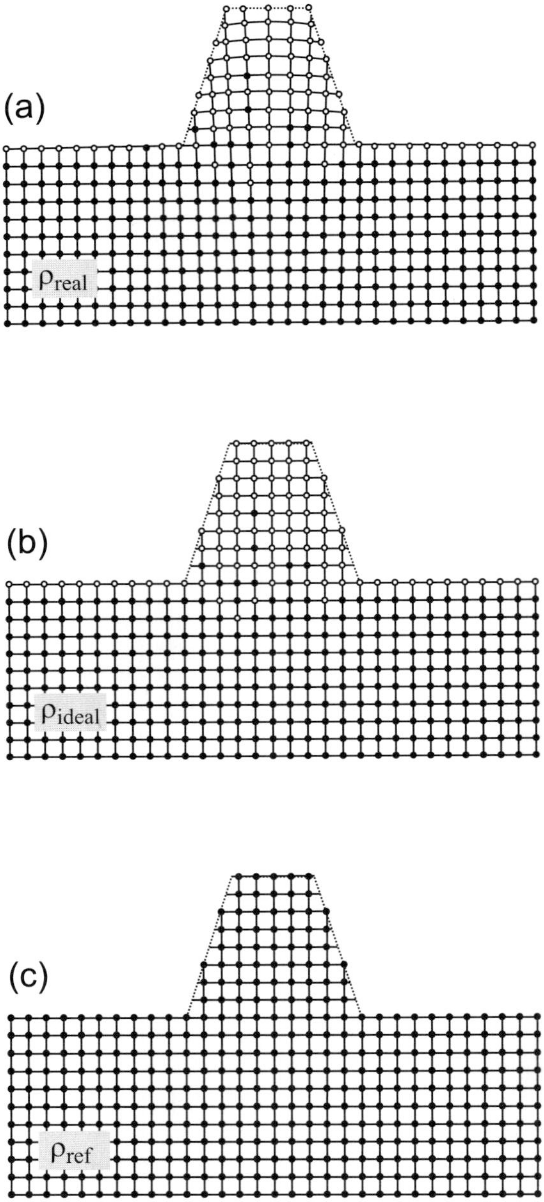

Fig. 2.5. (**a**) Real lattice $\varrho_{\text{real}}(\boldsymbol{r})$, (**b**) ideal lattice $\varrho_{\text{ideal}}(\boldsymbol{r})$, and (**c**) reference lattice $\varrho_{\text{ref}}(\boldsymbol{r})$, as used for the calculation of diffuse scattering from free-standing strained islands grown on a substrate. Different sorts of atoms are present, as indicated by *filled* and *open circles*. The reference lattice consists of a perfect periodic array of a single sort of atoms (*filled circles*). The ideal lattice is perfectly periodic concerning positions, however, different sorts of atoms are possible

parameter as the material of the mesoscopic structures in the elastically relaxed state. For that reason, the displacement field $\boldsymbol{u}(\boldsymbol{r})$ is referred to as the *total* displacement. Using

$$\int \varrho_{\text{ideal}}(\boldsymbol{r}-\boldsymbol{u}(\boldsymbol{r}))\mathrm{e}^{\mathrm{i}\boldsymbol{q}\cdot\boldsymbol{r}}\mathrm{d}V = \int \varrho_{\text{ideal}}(\boldsymbol{r})\mathrm{e}^{\mathrm{i}\boldsymbol{q}\cdot(\boldsymbol{r}+\boldsymbol{u}(\boldsymbol{r}))}\,\mathrm{d}V\,, \tag{2.39}$$

the diffuse amplitude can be written as

$$A^{\text{diffuse}}(\boldsymbol{q}) = A_0 r_0 P \frac{\mathrm{e}^{\mathrm{i}kR}}{R}\left[\int \varrho_{\text{ideal}}(\boldsymbol{r})\mathrm{e}^{\mathrm{i}\boldsymbol{q}\cdot(\boldsymbol{r}+\boldsymbol{u}(\boldsymbol{r}))}\,\mathrm{d}V \right. \\ \left. - \int \varrho_{\text{ref}}(z)\mathrm{e}^{\mathrm{i}\boldsymbol{q}\cdot\boldsymbol{r}}\,\mathrm{d}V\right]. \tag{2.40}$$

The second term on the right-hand side has only to be considered when one is calculating the diffuse intensity very close to a reciprocal-lattice point, i.e. when treating the diffuse scattering from the substrate. For the calculation of the diffuse scattering from strongly disturbed regions, e.g. the strain field inside free-standing islands, the second term on the right-hand side can be ignored.

We can express (2.40) in terms of the structure amplitude of the ideal crystal, which is defined as

$$F_{\boldsymbol{h}}^{\text{ideal}} = \int\limits_{\text{unit cell}} \varrho_{\text{ideal}}(\boldsymbol{r})\mathrm{e}^{\mathrm{i}\boldsymbol{h}\cdot\boldsymbol{r}_{\mathrm{B}}}\,\mathrm{d}V\,, \tag{2.41}$$

where $\boldsymbol{r}_{\mathrm{B}}$ is a vector inside the unit cell and $\boldsymbol{h}$ is an arbitrary reciprocal-lattice vector. The structure amplitude describes the scattering power of a single unit cell. Mesoscopic structures range from a few nanometers to about one micrometer in size. This means that the interesting features in reciprocal space are much smaller than the reciprocal unit vectors, i.e. we have to measure rather close to reciprocal-lattice points. Close to a reciprocal-lattice point, however, the scattering power of a unit cell is nearly constant and practically equal to that at the reciprocal-lattice point, i.e. equal to the structure amplitude. We assume that the variation $\Delta\boldsymbol{u}$ of the atomic displacements $\boldsymbol{u}$ within a unit cell is small, i.e. $|\boldsymbol{h}\cdot\Delta\boldsymbol{u}| \ll 1$. Strain fluctuations within the unit cells can then be neglected and the problem can be treated such that the entire unit cell is displaced by $\boldsymbol{u}(\boldsymbol{r})$ with respect to the ideal lattice. Using (2.38), we obtain

$$F_{\boldsymbol{h}}^{\text{real}} = F_{\boldsymbol{h}}^{\text{ideal}}\mathrm{e}^{\mathrm{i}\boldsymbol{q}\cdot\boldsymbol{u}(\boldsymbol{r})}\,. \tag{2.42}$$

For materials containing different sorts of atoms, e.g. binary, ternary, or quaternary systems, the atomic concentrations of the different species may vary inside the crystal. We take into account the local composition of the real sample at position $\boldsymbol{r}$ and choose a corresponding ideal crystal for that position. Note that the lattice parameter of the ideal lattice is always kept constant inside the entire sample; however, the local chemical composition can be taken into account. Therefore, the structure amplitude $F_{\boldsymbol{h}}^{\text{ideal}}$ of the ideal lattice depends on the position $\boldsymbol{r}$ inside the mesoscopic structure, and we can write

$$F_{\boldsymbol{h}}^{\text{real}} = F_{\boldsymbol{h}}^{\text{ideal}}(\boldsymbol{r})\mathrm{e}^{\mathrm{i}\boldsymbol{q}\cdot\boldsymbol{u}(\boldsymbol{r})} \, . \tag{2.43}$$

By using structure amplitudes, the integration in (2.40) can be replaced by a discrete sum over all unit cells k that take part in diffraction, and the diffusely scattered amplitude in the vicinity of the reciprocal-lattice point $\boldsymbol{h}$ can be written as

$$A_{\boldsymbol{h}}^{\text{diffuse}}(\boldsymbol{q}) = A_0 r_0 P \frac{\mathrm{e}^{\mathrm{i}kR}}{R} \sum_k \left(F_{\boldsymbol{h}}^{\text{ideal}}(\boldsymbol{r}_k)\mathrm{e}^{\mathrm{i}\boldsymbol{q}\cdot(\boldsymbol{r}_k+\boldsymbol{u}_k)} - F_{\boldsymbol{h}}^{\text{ref}}(z_k)\mathrm{e}^{\mathrm{i}\boldsymbol{q}\cdot\boldsymbol{r}_k} \right) , \tag{2.44}$$

where $\boldsymbol{r}_k$ is the position of the kth unit cell and $\boldsymbol{u}_k$ is its displacement. We can now calculate the differential scattering cross section, and obtain

$$\frac{\mathrm{d}\sigma}{\mathrm{d}\Omega}(\boldsymbol{q}) = r_0^2 P^2 S(\boldsymbol{q}) \, , \tag{2.45}$$

where

$$S(\boldsymbol{q}) = \left| \sum_k \left(F_{\boldsymbol{h}}^{\text{ideal}}(\boldsymbol{r}_k)\mathrm{e}^{\mathrm{i}\boldsymbol{q}\cdot(\boldsymbol{r}_k+\boldsymbol{u}_k)} - F_{\boldsymbol{h}}^{\text{ref}}(z_k)\mathrm{e}^{\mathrm{i}\boldsymbol{q}\cdot\boldsymbol{r}_k} \right) \right|^2 . \tag{2.46}$$

Finally, we have to stress that the ideal crystal is generally not identical to the reference lattice. For a multilayer of buried strained quantum dots, a laterally homogeneous perfect multilayer is chosen as the reference lattice. For free-standing islands, the unstrained (homogeneous) substrate lattice can be chosen as the reference lattice (Fig. 2.5). Equation (2.44) thus applies to the most general case of calculating the diffuse scattering from a (slightly) distorted system in the framework of kinematical scattering theory.

On the basis of (2.44), various mechanisms of intensity contrast are possible:

- The scattering power depends on the atomic species inside the mesoscopic structure. Therefore, a change or gradient in the atomic composition (for binary, ternary, quaternary, etc. systems) may lead to a considerable fluctuation of the structure amplitude $F_{\boldsymbol{h}}^{\text{ideal}}(\boldsymbol{r}_k)$. These fluctuations, which are generally related to the growth process (e.g. interdiffusion or segregation), generate *structure factor contrast*, which is very similar to the absorption contrast in X-ray tomography or electron microscopy.
- Intensity contrast can be induced by strain inside the mesoscopic structure. Since the strain enters into the diffuse intensity via the phase $\boldsymbol{q}\cdot(\boldsymbol{r}_k + \boldsymbol{u}_k)$, the term *phase contrast* is often used.
- The summation in (2.44) is performed over the entire mesoscopic structure, e.g. a nanoscale island with a limited size and a distinct shape. Therefore, truncation effects are visible in the intensity distribution.

Among these effects are crystal truncation rods and thickness fringes, originally known as Kiessig fringes [56]. These shape- and size-related features in diffuse scattering determine the *form function* in reciprocal space.

Usually, it is not straightforward to distinguish between these three effects. Moreover, strain, shape, size, and chemical composition cannot be directly extracted from the intensity distribution in reciprocal space. The diffuse-scattering intensity pattern has to be calculated by assuming a certain structure model in real space, and this model has to be refined until satisfactory agreement between simulation and experiment is achieved. How this is performed will be described in more detail in the following section.

2.3.3 Numerical X-Ray Simulations Using the Finite-Element Method

For several reasons it is impossible – at least in general – to directly retrieve the required information from X-ray diffuse scattering. Some of these reasons are the loss of phase information in intensity measurement, the superimposition of effects of strain, shape, and positional correlation, and the tensor character of strain. Researchers have tried to overcome the latter point by using analytical expressions to approximate the strain field, ensuring rapid calculations of strain. These analytical approaches have been applied for planar layers with steps [57], quantum wires [58], hut clusters [59], and rotationally symmetric islands [60]. They are, however, often empirical [58, 59] and, owing to the complicated strain tensor field $\varepsilon_{ij}(x, y, z)$ of a three-dimensional island, not justified for complicated three-dimensional structures.

Often, the Green's function approach is used, and full analytical expressions for the strain distribution have been derived for ellipsoidal, cuboidal, and truncated-pyramidal dots [61, 62]. However, the main disadvantages of the Green's function approach are a restriction to buried structures and the assumption of elastic isotropy, which is just a crude approximation for cubic semiconductors. Therefore, the simple isotropic approach may miss some key features of the strain field. However, recently a Green's tensor method has been developed which enables the strain distribution in and around quantum dots to be calculated including the anisotropy of the elastic constants [63].

Kegel et al. [64, 65, 66] have developed an interesting approach called *iso-strain scattering*, which can be applied to islands with cylindrical symmetry. These authors make use of a four-channel scattering process (see Sect. 2.5.2) but omit scattering from the substrate.

In recent years, the use of the finite-element method (FEM) to calculate the strain field in mesoscopic structures has become successfully established (e.g. [67, 68, 69, 70]). The FEM is based on linear continuum elasticity theory, it includes the full elastic anisotropy, and it has proven to be applicable down to structure sizes of a few tens of nanometers. At object sizes below about

20 nm and at very high lattice mismatch (e.g. 7% for InAs quantum dots on GaAs), there are distinct deviations from elasticity theory and the atomic structure of the quantum dots has to be considered [71]. However, these atomistic calculations are mostly performed in the framework of the valence force field (VFF) method. Even with high-performance computers, true first-principles (ab initio) calculations are feasible only for systems containing fewer than about 100 atoms [72, 73].

For the systems under consideration here, linear elasticity theory is valid. Wiebach et al. [68] have developed the following brute force iterative approach for data evaluation of nanoscale islands:

1. Creation of a specific structure model in real space that includes island size, shape, and chemical composition.
2. FEM calculation of the three-dimensional strain field inside the island and the surrounding substrate and barrier layer.
3. Numerical brute force calculation of diffuse scattering using the formula described in the previous section (2.44).
4. Comparison with experimental data.
5. Further improvement of the model (1) and calculation of diffuse intensities from the new model until satisfactory agreement is achieved.

The components of the displacement field $\boldsymbol{u}(\boldsymbol{r}) = (u_1, u_2, u_3)$ that is used in the X-ray simulations are related to the (symmetrical) elastic strain tensor components ε_{ij} through

$$\varepsilon_{ij} = \frac{1}{2}\left(\frac{\partial u_i}{\partial x_j} + \frac{\partial u_j}{\partial x_i}\right); \quad i, j = 1, 2, 3\,. \tag{2.47}$$

The *total* strain is employed, i.e. the lattice parameter of the underlying substrate is used as the reference. This is advantageous because the atomic displacement has been chosen with respect to the ideal lattice (of the substrate). For more detailed information see Sect. A.2.

In general, this approach cannot be used as a true "fitting" procedure since there are too many free parameters in the model (shape, size, chemical composition, and spatial correlation). Therefore, it is necessary to include knowledge obtained by other methods, e.g. information on the shape and size obtained by AFM, SEM, and TEM. First, the calculation is done for a single island and its environment. The subsequent simulation of the diffuse intensity is performed numerically, i.e. on a regular grid consisting of *base cells*. If necessary, the dimensions of the base cells can be chosen as small as that of the crystal unit cells; however, they can also be chosen larger. The numerical procedure has to be carried out for some 100 000 to 1 million base cells and more than 10 000 values of q, and is therefore rather time-consuming. Positional correlation then has to be incorporated. This will be discussed in more detail in the next subsection.

We now present some results of such calculations where the influence of strain, shape, and chemical composition are explicitly demonstrated. Let us consider a truncated pyramid (base width $w = 30\,\mathrm{nm}$, height $h = 18\,\mathrm{nm}$, $\{111\}$ side facets) grown on a silicon substrate, and let us investigate the diffuse intensity in the vicinity of the 004 reciprocal-lattice point. First, we "switch off" the strain and the structure factor contrast by calculating the hypothetical case of a homogeneous Si island grown on the Si substrate. In this case only the third item of the list given at the end of Sect. 2.3.2 has to be taken into account. The resulting shape and truncation effects in the intensity distribution can be inspected in Fig. 2.6a. These consist of (i) thickness fringes that reflect the finite dimensions of the island, and (ii) crystal truncation

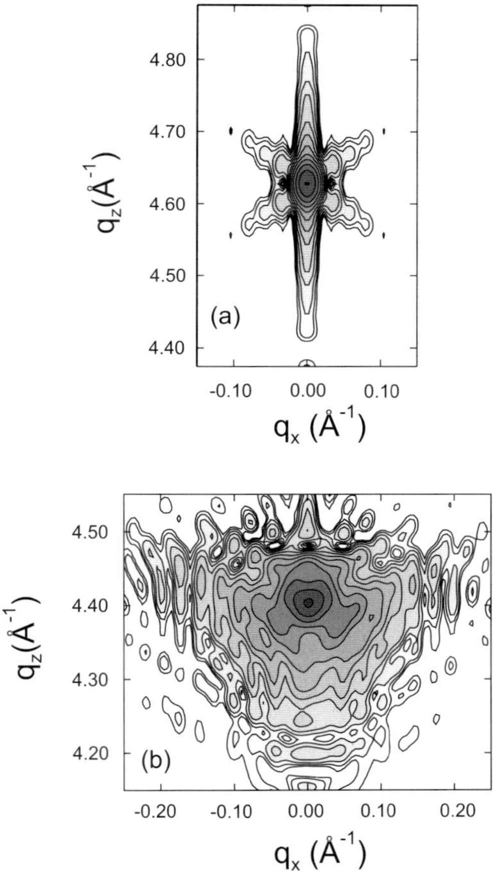

Fig. 2.6. Calculated diffuse intensity in the vicinity of the 004 reciprocal-lattice point for a truncated pyramid ($w = 30\,\mathrm{nm}$, $h = 18\,\mathrm{nm}$): (**a**) no strain, (**b**) strained Ge island on Si substrate (lattice mismatch 4.2%). Adjacent contour levels differ by a factor of $\sqrt{2}$. The strain profile was calculated by the finite-element method [68]

rods (CTRs) that are generated by the {111} side facets. When the strain is "switched on" by replacing the Si inside the island by pure germanium (lattice mismatch 4.2%), the intensity distribution is drastically changed into a butterfly-shaped diffraction pattern, which, however, exhibits a remaining influence of thickness fringes (Fig. 2.6b).

This example proves that the intensity distribution is dominated mostly by strain and partly by shape effects. On the other hand, the structure factor contrast mentioned above usually plays a negligible role [74].

2.3.4 Influence of Positional Correlation

The investigation of mesoscopic structures by means of X-ray diffuse scattering implies averaging over the illuminating spot size of typically a few square millimeters on the sample. When a single layer of islands with a spacing of 100 nm–1000 nm distance is considered, the number of scattering mesoscopic objects is of order 10^6–10^8, which is a statistically significant large number. In the case of such a large ensemble of mesoscopic structures, it is sufficient to focus on the diffraction from a single island if the islands are monodisperse. In this case the diffuse intensity from each island is identical. However, the positions of these islands in the ensemble can be correlated, resulting in significant interference of the diffusely scattered waves from the individual islands. Depending on the strength of the positional correlation, this could lead to more or less pronounced and sharp satellite peaks in the diffuse scattering. The distances Δq in reciprocal space between the satellites and the coherent Bragg peak are related to corresponding mean distances $\langle d \rangle$ in real space via

$$\langle d \rangle = \frac{2\pi}{\Delta q} . \tag{2.48}$$

This equation can be applied to both vertical and horizontal correlation and is very helpful for quickly evaluating experimental data. However, this relationship does not provide information about the details of the autocorrelation function that determines the scattered intensities. Here, we describe how to take account of positional correlation in the framework of kinematical theory. The diffraction from an ensemble of nanoscale islands can be treated such that the amplitudes A_m^{diffuse} from islands at positions $\boldsymbol{R}_m$ are (coherently) summed. This approximation is valid as long as the distances between the islands are significantly larger than their size. In this case we can write the following for the totally scattered amplitude:

$$\boldsymbol{A}_{\text{total}}^{\text{diffuse}}(\boldsymbol{q}) = \sum_m \boldsymbol{A}_m^{\text{diffuse}}(\boldsymbol{q}) \mathrm{e}^{\mathrm{i}\boldsymbol{q}\cdot\boldsymbol{R}_m} . \tag{2.49}$$

For practical reasons, it is impossible to extract useful information about the individual objects when they are different in size, shape, and chemical composition. Extracting information is only possible if all of them are sufficiently similar, i.e. they must have the same shape (including orientation) and at least a narrow size distribution. In the case of identical islands at positions $\boldsymbol{R}_m$, the amplitudes $\boldsymbol{A}_m^{\text{diffuse}}(\boldsymbol{q})$ are also identical and (2.49) transforms into

$$\boldsymbol{A}_{\text{total}}^{\text{diffuse}}(\boldsymbol{q}) = \boldsymbol{A}^{\text{diffuse}}(\boldsymbol{q}) \sum_m \mathrm{e}^{\mathrm{i}\boldsymbol{q}\cdot\boldsymbol{R}_m} . \tag{2.50}$$

At first sight the assumption of identical islands seems to be a rather drastic and limiting requirement. However, this coincides with the requirements of possible device applications, especially in the case of quantum dots. Indeed, for many practical systems this is a sufficiently good approximation. The total diffuse intensity can be then written as

$$\left(\frac{\mathrm{d}\sigma}{\mathrm{d}\Omega}\right)_{\text{ensemble}}(\boldsymbol{q}) = \frac{\mathrm{d}\sigma}{\mathrm{d}\Omega}(\boldsymbol{q})G(\boldsymbol{q}) , \tag{2.51}$$

where the interference function is

$$G(\boldsymbol{q}) = \left|\sum_m \mathrm{e}^{\mathrm{i}\boldsymbol{q}\cdot\boldsymbol{R}_m}\right|^2 = \sum_{m,n} \mathrm{e}^{\mathrm{i}\boldsymbol{q}\cdot(\boldsymbol{R}_m-\boldsymbol{R}_n)} . \tag{2.52}$$

In (2.52), the summation is performed over the entire ensemble of mesoscopic structures. This requires a knowledge of all island positions $\boldsymbol{R}_m$, which is practically impossible. Equation (2.52) should therefore read

$$G(\boldsymbol{q}) = \left\langle \sum_{m,n} \mathrm{e}^{\mathrm{i}\boldsymbol{q}\cdot(\boldsymbol{R}_m-\boldsymbol{R}_n)} \right\rangle , \tag{2.53}$$

where the brackets $\langle \ldots \rangle$ denote averaging over all possible configurations of the mesoscopic ensemble. To a certain extent the positions $\boldsymbol{R}_m$ can be determined by imaging techniques such as atomic force microscopy (AFM). Here, a sufficiently large area in real space has to be considered. However, the result obtained has to be adapted to the experimental conditions, i.e. the finite coherence of the X-ray beam has to be taken into account. This means that the coherent sum in (2.52) has to be taken over those islands that are inside the coherence volume. This sum has then to be averaged over all islands. In other words, the total intensity is calculated by partly coherent and partly incoherent averaging over different subensembles. An example where this averaging is performed will be discussed later.

Equation (2.51) shows that, as a result of spatial correlation, the diffuse intensity of a single island is modulated by the interference function $G(\boldsymbol{q})$. Note that $G(\boldsymbol{q})$ is defined in reciprocal space, whereas the autocorrelation function $C(\boldsymbol{r})$ given in (2.35) is defined in real space. There exists, however, an interesting relationship between $C(\boldsymbol{r})$ and $G(\boldsymbol{q})$, which can be obtained by expressing the positions $\boldsymbol{R}_m$ of the islands in terms of the "island density" through

$$\varrho(\boldsymbol{r}) = \varrho_0 \sum_m \delta(\boldsymbol{r} - \boldsymbol{R}_m) \; . \tag{2.54}$$

Here, the extended size of the mesoscopic structures is neglected and only the mean positions $\boldsymbol{R}_m$ are considered. By using this relationship for calculating the autocorrelation function $C(\boldsymbol{r})$ via (2.35), we obtain

$$G(\boldsymbol{q}) \propto \int_{\text{ensemble}} C(\boldsymbol{r}) \mathrm{e}^{\mathrm{i}\boldsymbol{q}\cdot\boldsymbol{r}} \, \mathrm{d}V \; , \tag{2.55}$$

which is rather similar to (2.34). The interference function in reciprocal space is the Fourier transform of the autocorrelation function in real space, and vice versa.

Ordering phenomena are usually discussed in the framework of two limiting models – *short-range order* (SRO) and *long-range order* (LRO). For LRO, the correlation is assumed to be perfect within coherent domains of size ξ, i.e. $C(\boldsymbol{r})$ exhibits (sharp) peaks of constant width independent of $\boldsymbol{r}$. In this case $G(\boldsymbol{q})$ also shows correlation peaks with constant widths, which are given by Scherrer's formula [75],

$$\delta q = \frac{2\pi}{\xi} \; . \tag{2.56}$$

The domain size ξ can then be interpreted as a correlation length which quantitatively describes the length scale up to which positional correlation is present.

For SRO, the ordering disappears gradually at large distances. Therefore the peaks in $C(\boldsymbol{r})$ become broader and are damped with increasing relative distance r. For the rms peak widths of $C(\boldsymbol{r})$, Stangl et al. [76] have assumed that they behave similarly to a random walk, i.e.

$$\sigma_n = \sqrt{n}\sigma \; . \tag{2.57}$$

Here σ is the rms standard deviation of the mean distance $\langle d \rangle$ between two adjacent islands and is, thus, identical to the peak width of the first correlation peak of $C(\boldsymbol{r})$. As a consequence of (2.57), the peak widths of $G(\boldsymbol{q})$ also increase with the peak index. Stangl et al. [76] have given the result

$$\delta q = \frac{(\sigma q)^2}{\langle d \rangle} \; , \tag{2.58}$$

with a correlation length

$$\xi = \frac{\langle d \rangle^3}{2\sigma^2} \; . \tag{2.59}$$

Comparison of (2.58) and (2.59) then leads to an expression for the width of the first correlation peak,

$$\delta q = \frac{2\pi^2}{\xi} \,, \tag{2.60}$$

which differs by a factor of π from (2.56) for LRO. It is, however, well known that the relationship between peak width and correlation length depends on the kind of autocorrelation function used, and the corresponding values of ξ are often rather different.

Stangl et al. [76] have proven the validity of the SRO model for SiGe and PbSe/PbEuTe quantum dot superlattices. For the latter case, experimental data and a quadratic fit to the experimental peak widths are presented in Fig. 2.7. The degree of ordering is remarkably high, but the SRO behavior is still evident.

The qualitative difference between SRO and LRO is sketched in Fig. 2.8. In both models the peaks in $G(\boldsymbol{q})$ decay for larger q and are equally spaced; however, the peak widths are constant for LRO, whereas they become larger with increasing q for the SRO model.

For SRO as described by (2.57), an analytical expression for the interference function $G(\boldsymbol{q})$ exists and is given by

$$G(\boldsymbol{q}_\parallel) = \frac{1 - \mathrm{e}^{-\sigma^2 q_\parallel^2}}{1 + \mathrm{e}^{-\sigma^2 q_\parallel^2} - 2\mathrm{e}^{-\sigma^2 q_\parallel^2/2}\cos(q_\parallel \langle d \rangle)} \,. \tag{2.61}$$

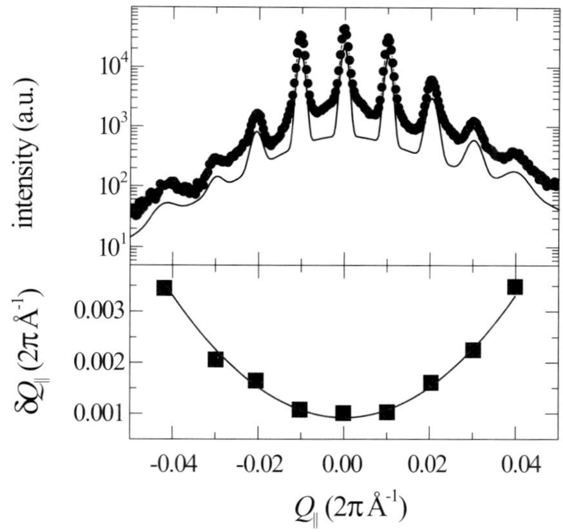

Fig. 2.7. GISAXS scan (see Sect. 3.6.3) of a PbSe/PbEuTe quantum dot superlattice sample (*upper panel*) in an azimuth $q_\parallel$ parallel to $\langle 1\bar{1}0 \rangle$. The lateral widths of the maxima are displayed in the *lower panel*. The lateral correlation properties in the SRO model were obtained from a quadratic fit in accordance with (2.58). The figure has been taken from Stangl et al. [76]

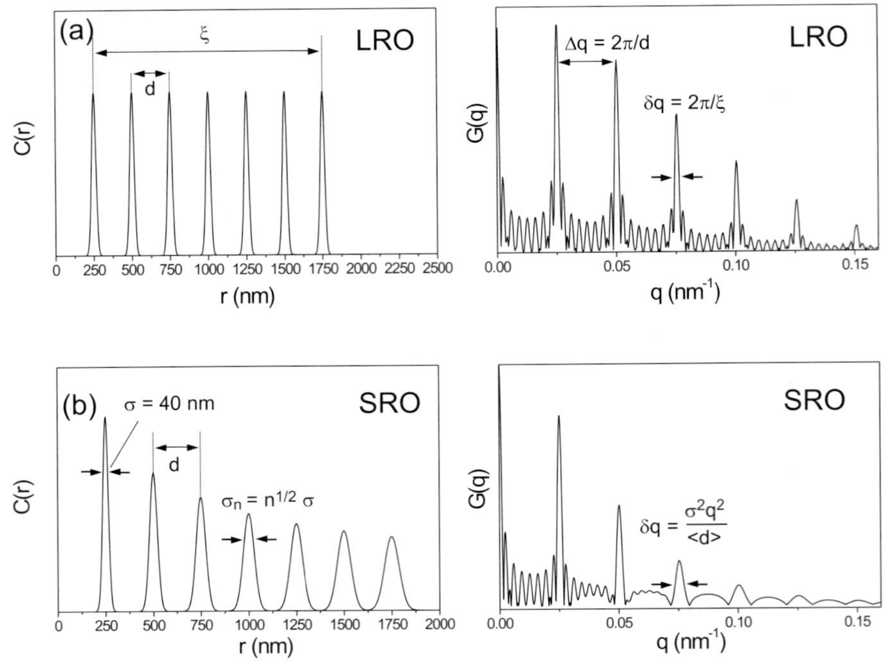

Fig. 2.8. Autocorrelation function $C(\boldsymbol{r})$ in real space and the corresponding interference function $G(\boldsymbol{q})$ in reciprocal space for (**a**) long-range ordering (LRO), and (**b**) short-range ordering (SRO) using a random-walk-like behavior. For LRO the peak widths and peak spacings of $C(\boldsymbol{r})$ and $G(\boldsymbol{q})$ are constant, whereas for SRO the peak widths increase with the peak index. For details see text

This expression was initially developed to describe scattering from random stacking faults without an average lattice [4]. Metzger et al. [77, 78] have used it for treating in-plane correlation of free-standing quantum dots. Here $\langle d \rangle$ and σ were used as free fitting parameters. They were treated as anisotropic in real space, since a generalized two-dimensional (in q-space) analytical expression for $G(\boldsymbol{q})$ does not exist yet.

A very interesting point is the relationship between $C(\boldsymbol{r})$ and $G(\boldsymbol{q})$. What qualitative information can be obtained from the functional behavior of $G(\boldsymbol{q})$? In the literature (e.g. [79]), the statement can often be found that the presence of many peaks in $G(\boldsymbol{q})$, i.e. the presence of high-order correlation peaks, can be interpreted as very high correlation. This statement is, however, not generally correct, and will be discussed now in more detail. It is certainly true that the degree of ordering is related to the peak width of the first correlation peak, and the corresponding correlation length is given by (2.56) or (2.60), depending on whether the LRO or SRO model is applied. This behavior can be inspected in Fig. 2.9, where calculated values of the interference function

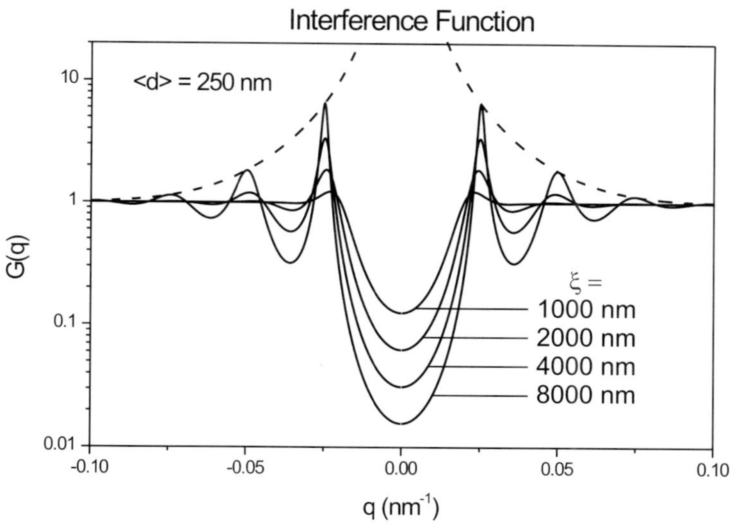

Fig. 2.9. Calculated interference function $G(\boldsymbol{q}_{\|})$ obtained using (2.61) for $\langle d \rangle = 250\,\mathrm{nm}$ and various values of the correlation length ξ. The *dashed line* is the envelope function of the satellite peaks for $\xi = 8000\,\mathrm{nm}$

$G(\boldsymbol{q}_{\|})$ given by (2.61) are plotted for different values of ξ. The width of the first order correlation peak indeed inversely scales with the correlation length ξ in accordance with (2.60). In addition to that behavior, the correlation peaks decay with increasing values of q. The intensity decay can be quantified by an envelope function of the form $(1+\mathrm{e}^{-\sigma^2 q^2/2})/(1-\mathrm{e}^{-\sigma^2 q^2/2})$. The smaller ξ is, the stronger the decay of the correlation peaks. In the SRO model as used in (2.57), the appearance of many orders of satellite peaks is thus a sign of high correlation. This behavior, however, is not generally valid. In Fig. 2.10, an SRO model where only next-neighbor interaction is present is used. When an asymmetric peak profile is used for $C(\boldsymbol{r})$ instead of a symmetrical Gaussian distribution, interesting features in the resulting intensity distribution $G(\boldsymbol{q})$ are obtained:

- Many orders of correlation peaks are present, although only poor SRO is present.
- The position of the first order correlation peak in $G(\boldsymbol{q})$ is given by $q = 2\pi/\langle d \rangle$; however, the spacings between neighboring peaks disperse slightly outwards. This is a direct consequence of the asymmetric peak profile of $C(\boldsymbol{r})$.
- The correlation peaks in $G(\boldsymbol{q})$ become slightly broader with increasing values of $\boldsymbol{q}$. However, a quadratic behavior, as given by the random-walk-like SRO model (2.58), is not observed.

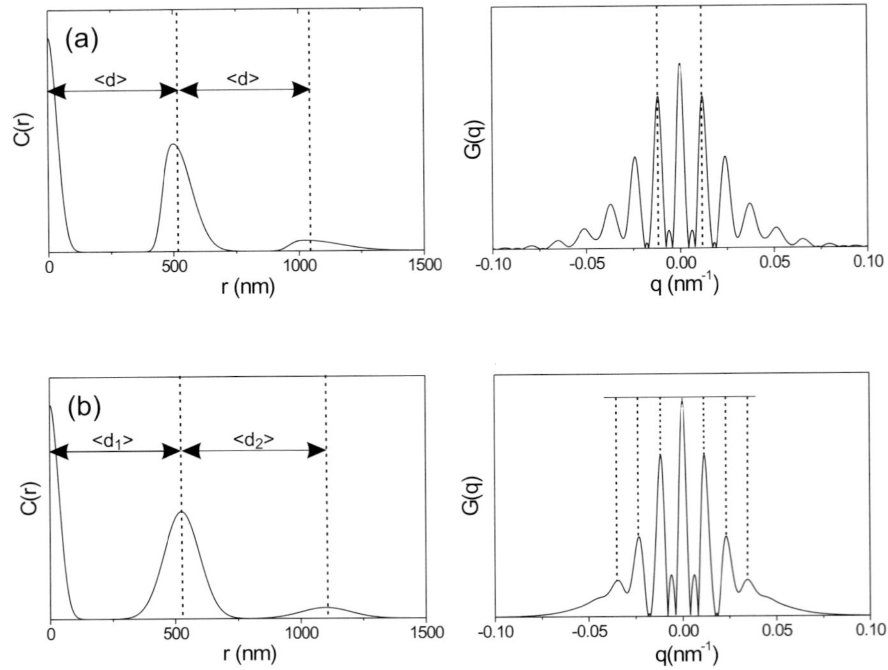

Fig. 2.10. Calculated interference and autocorrelation functions $G(\boldsymbol{q}_{\|})$ and $C(\boldsymbol{r})$, for two special SRO models, both exhibiting only next-neighbor correlations. In (**a**), the line shape in $C(\boldsymbol{r})$ is asymmetric. As a consequence, many high-order correlation peaks are present in $G(\boldsymbol{q})$ that are not equidistantly spaced but disperse slightly outwards with increasing values of q. In (**b**), the second-order peak of $C(\boldsymbol{r})$ has been chosen such that it appears at slightly higher values of r, while the line profiles are symmetrical. As a consequence, the corresponding correlation peaks in $G(q)$ are strongly damped, while the peaks are still equidistantly spaced

The occurrence of nonequally spaced correlation peaks in the diffuse scattering cannot be explained by varying peak spacings in real space. This would only lead to a stronger damping of the intensity distribution with increasing q, while the correlation peaks would be exactly equidistant. This is also illustrated in Fig. 2.10.

In this book, both free-standing mesoscopic structures and buried structures stacked in a multilayer are considered. In the case of independent vertical and horizontal correlation, $G(\boldsymbol{q})$ can be factorized into vertical and horizontal components:

$$G(\boldsymbol{q}) = G_{\perp}(q_z)G_{\|}(\boldsymbol{q}_{\|}) . \tag{2.62}$$

The vertical interference function describes vertical inheritance in the mesoscopic structures, whereas the horizontal interference function is identical for each layer of mesoscopic structures and is thus independent of the layer index. This approximation is not generally valid. On one hand, it has been shown that growth of subsequent layers of quantum dots may lead to an improved size homogeneity and increasing horizontal ordering [80, 81, 82]. On the other hand, inclined (oblique or skew) [83, 84, 85] or even more complicated vertical correlation [86, 87] may also be present. In this case (2.62) has to be modified. However, the expression given in (2.62) simplifies the theoretical treatment of vertical correlation.

Kegel et al. [88] have used a random-walk model in order to investigate the vertical alignment of laterally nonordered quantum dots as a function of the vertical layer index. Using this model, they have derived an analytical expression for the $q_\perp$ width of diffuse scattering,

$$\delta q_\perp = \frac{\sigma^2}{\langle d \rangle} q_\parallel^2 + \frac{\mu}{\langle d \rangle} , \tag{2.63}$$

where μ is an attenuation factor which describes the loss of intensity in the X-ray beam from one bilayer to the next. Their experimental results obtained from multilayered Ge/Si(001) quantum dots are in agreement with this model (see Fig. 2.11).

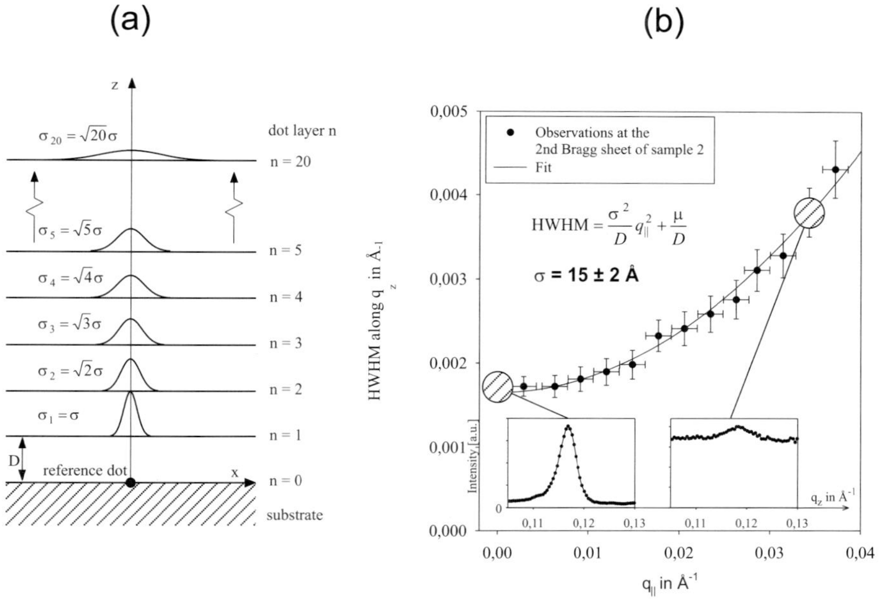

Fig. 2.11. (**a**) Model for the lateral deviations from an average dot position. The half-width of the spread in lateral position increases in a random-walk-like manner. (**b**) Experimental results for the half-width of diffuse scattering from a Ge/Si quantum dot superlattice. For more details see [88]

2.3.5 Small-Angle X-Ray Scattering

Equation (2.37) describes the diffuse intensity in the vicinity of an arbitrary reciprocal-lattice point. A special case is the diffuse intensity in the vicinity of $\boldsymbol{h} = 0$. This scattering geometry is referred to as small-angle X-ray scattering (SAXS). In the context of SAXS, we always average over the electron density within a unit cell, i.e. the discrete atomic structure inside the unit cell is neglected. In this case (2.44) can be written as a sum over all unit cells:

$$A_{\mathrm{SAXS}}(\boldsymbol{q}) = A_0 r_0 P \frac{\mathrm{e}^{\mathrm{i}kR}}{R} \sum_k \Bigg(\underbrace{\left(F_0^{\mathrm{ideal}}(\boldsymbol{r}_k) - F_0^{\mathrm{ref}}(z_k)\right)}_{=\Delta F_0(\boldsymbol{r}_k)} \Bigg) \mathrm{e}^{\mathrm{i}\boldsymbol{q}\cdot\boldsymbol{r}_k} \,. \tag{2.64}$$

It can be easily seen that there is no influence of atomic distortion on the SAXS signal. The amplitude of the scattered wave is sensitive to the coherent sum of the difference $\Delta F_0(\boldsymbol{r}_k)$ in the structure amplitudes between the ideal and reference lattices. $\Delta F_0(\boldsymbol{r}_k)$ changes significantly only with spatial changes in the electron density. Therefore, SAXS probes electron density fluctuations.

The most common case, of homogeneous mesoscopic structures with different sizes and shapes, can be expressed through the electron density given by

$$\varrho(\boldsymbol{r}) = \varrho_0 \sum_m \Omega_m(\boldsymbol{r} - \boldsymbol{R}_m) \,, \tag{2.65}$$

where $\Omega_m(\boldsymbol{r} - \boldsymbol{R}_m)$ is the shape function of island m at position $\boldsymbol{R}_m$, which is defined as

$$\Omega_m(\boldsymbol{r}) = \begin{cases} 1 \text{ inside island} \\ 0 \text{ outside island} \end{cases} . \tag{2.66}$$

For monodisperse structures with identical shapes ($\Omega_m(\boldsymbol{r} - \boldsymbol{R}_m) = \Omega(\boldsymbol{r} - \boldsymbol{R}_m)$), (2.65) can be simplified to

$$\varrho(\boldsymbol{r}) = \varrho_0 \sum_m \Omega(\boldsymbol{r} - \boldsymbol{R}_m) = \varrho_0 \Omega(\boldsymbol{r}) \otimes \sum_m \delta(\boldsymbol{r} - \boldsymbol{R}_m) \,, \tag{2.67}$$

where $\otimes$ is the convolution product.

A very simple electron density fluctuation is already present for a single, homogeneous, free-standing island. In that case we can write the following for the electron density for an island at position $\boldsymbol{R}_m$:

$$\varrho(\boldsymbol{r}) = \varrho_0 \Omega(\boldsymbol{r} - \boldsymbol{R}_m) \,. \tag{2.68}$$

Consequently, $F_0^{\mathrm{ideal}}(\boldsymbol{r}_k) = F_0 = \text{const}$ inside the island, $F_0^{\mathrm{ideal}}(\boldsymbol{r}_k) = 0$ outside the island, and – when only scattering from the island is considered – $F_0^{\mathrm{ref}} = 0$. The scattered intensity is then given by

$$\left(\frac{\mathrm{d}\sigma}{\mathrm{d}\Omega}\right)_{\mathrm{single}}(\boldsymbol{q}) = r_0^2 P^2 \varrho_0^2 |\Omega^{\mathrm{FT}}(\boldsymbol{q})|^2 \,, \tag{2.69}$$

where the island form factor $\Omega^{\mathrm{FT}}(\boldsymbol{q})$ is the Fourier transform of the shape function $\Omega(\boldsymbol{r}-\boldsymbol{R}_m)$.

Using (2.65), the total scattered amplitude for an ensemble of identical structures can be expressed as

$$A(\boldsymbol{q}) = A_0 r_0 P \varrho_0 \frac{\mathrm{e}^{\mathrm{i}kR}}{R} \Omega^{\mathrm{FT}}(\boldsymbol{q}) \sum_m \mathrm{e}^{\mathrm{i}\boldsymbol{q}\cdot\boldsymbol{R}_m} , \tag{2.70}$$

and, similarly to (2.51), the total intensity is given by

$$\left(\frac{\mathrm{d}\sigma}{\mathrm{d}\Omega}\right)_{\text{ensemble}}(\boldsymbol{q}) = r_0^2 P^2 \varrho_0^2 |\Omega^{\mathrm{FT}}(\boldsymbol{q})|^2 G(\boldsymbol{q}) . \tag{2.71}$$

This equation is in complete equivalence to (2.51), where spatial ordering is treated by means of the interference function $G(\boldsymbol{q})$. A very important characteristic of (2.71) is that the total intensity can be factorized into shape-induced ($\Omega^{\mathrm{FT}}(\boldsymbol{q})$) and correlation-induced ($G(\boldsymbol{q})$) diffuse scattering. This relationship holds, however, only for mesoscopic ensembles in which the mean distance between the mesoscopic structures is significantly larger than their lateral sizes. In this case both factors in (2.71) are spatially separated in reciprocal space, and shape- and correlation-induced diffuse scattering can be distinguished.

Another difference between shape- and correlation-induced scattering is their symmetry. Since an island is a three-dimensional object, the island form factor $\Omega^{\mathrm{FT}}(\boldsymbol{q})$ is also three-dimensional in reciprocal space. The expression for $G(\boldsymbol{q})$ in (2.52) holds for an arbitrary three-dimensional ensemble of mesoscopic structures. In the case of a single layer, only lateral correlation has to be considered. In that case the correlation function depends only on the in-plane component of the scattering vector, $\boldsymbol{q}_\|$, i.e. $G(\boldsymbol{q}) = G(\boldsymbol{q}_\|)$. Consequently, lateral positional correlation leads to a rodlike structure of diffuse intensity perpendicular to the sample surface and a pronounced in-plane intensity modulation.

For free-standing structures, the in-plane intensity distribution of diffuse scattering in the vicinity of $\boldsymbol{h} = 0$ can be directly compared with the in-plane power spectrum $P(q_x, q_y)$ of the height profile $h(x, y)$, as measured by atomic force microscopy, for example. This power spectrum is defined as

$$P(q_x, q_y) = \left| \int_A h(x,y) \mathrm{e}^{\mathrm{i}(q_x x + q_y y)} \, \mathrm{d}x \, \mathrm{d}y \right|^2 . \tag{2.72}$$

Here, the integration is performed over the measured surface area A. Neglecting refraction effects, this quantity is basically identical to the in-plane intensity distribution given in (2.71), as long as A is chosen sufficiently large [89].

2.4 Dynamical Scattering Theory

In the previous sections, only kinematical scattering has been considered. This approach is very powerful for strongly disturbed systems but is often not sufficient for the theoretical treatment of rather perfect semiconductor systems as discussed here. One reason is the high crystalline perfection of these systems. This may lead – under certain scattering conditions – to strong multiple-diffraction phenomena. The other reason is that the mesoscopic structures are grown on a thick substrate. In order to increase sensitivity, a grazing-incidence and/or grazing-exit scattering geometry is often chosen (see Sects. 3.6.2 and 3.6.3). In this case refraction effects, as discussed in Sect. 2.2.2, have to be considered.

On the other hand, the dynamical diffraction theory [1, 2] can be applied to perfect systems, but it is not able to treat diffuse scattering from the mesoscopic structures discussed in the present work. The reason why dynamical diffraction is relevant is the fact that perturbation theory – which will be described in the next section – uses the exact dynamical solutions of an undisturbed system as input for further calculations. An understanding of the main aspects of dynamical theory is thus important for anyone who wishes to read further.

Although the X-ray wavelengths considered here are in the range of interatomic distances, i.e. in the angstrom range, many aspects of the interaction of hard X-rays with matter can be described by classical electrodynamics. For the most common case of elastic interaction, the wave inside the crystal cannot be simply expressed as a set of single monochromatic plane waves (see (2.11)), but has generally to be written as

$$\boldsymbol{E}(\boldsymbol{r},t) = \boldsymbol{E}(\boldsymbol{r})\mathrm{e}^{-\mathrm{i}\omega t} \; . \tag{2.73}$$

The wave fields $\boldsymbol{E}(\boldsymbol{r})$ inside and outside the sample are then the solutions of the stationary wave equation as derived from Maxwell's equations,

$$(\Delta + k^2)\boldsymbol{E}(\boldsymbol{r}) = \text{grad div } \boldsymbol{E}(\boldsymbol{r}) - k^2\chi(\boldsymbol{r})\boldsymbol{E}(\boldsymbol{r}) \; , \tag{2.74}$$

where $k = \omega/c_0$ is the length of the vacuum wave vector and $\chi(\boldsymbol{r})$ is the complex dielectric polarizability given in (2.1). If we define the *scattering potential operator* $\hat{\mathbf{V}}$ through

$$\hat{\mathbf{V}}\boldsymbol{E}(\boldsymbol{r}) \equiv \text{grad div } \boldsymbol{E}(\boldsymbol{r}) - k^2\chi(\boldsymbol{r})\boldsymbol{E}(\boldsymbol{r}) \; , \tag{2.75}$$

we obtain

$$(\Delta + k^2)\boldsymbol{E}(\boldsymbol{r}) = \hat{\mathbf{V}}(\boldsymbol{r})\boldsymbol{E}(\boldsymbol{r}) \; . \tag{2.76}$$

This stationary wave equation – also referred to as the Helmholtz equation – is the most general description of elastic scattering. It can be simplified at small scattering angles [13] by using the approximation[7]

$$\operatorname{div} \boldsymbol{E}(\boldsymbol{r}) = \frac{1}{\varepsilon_0} \operatorname{div}\left(\frac{\boldsymbol{D}(\boldsymbol{r})}{1+\chi(\boldsymbol{r})}\right) \approx \frac{1}{\varepsilon_0(1+\chi_0)} \operatorname{div} \boldsymbol{D}(\boldsymbol{r}) = 0 , \tag{2.77}$$

where $\boldsymbol{D}(\boldsymbol{r})$ is the electric induction vector. In this case the "grad div" term in (2.75) vanishes and we can write (if we neglect absorption)

$$\hat{\mathbf{V}} = V(\boldsymbol{r}) = -k^2 \chi_0(\boldsymbol{r}) = 4\pi r_0 \varrho(\boldsymbol{r}) . \tag{2.78}$$

As can be seen from this relationship, the scattering potential is a linear function of the polarizability $\chi_0(\boldsymbol{r})$ and, consequently, of the electron density $\varrho(\boldsymbol{r})$.

For a perfect crystal, a useful relationship between $\chi(\boldsymbol{r})$ and the structure amplitude $F_{\boldsymbol{h}}$ can be found by expanding $\chi(\boldsymbol{r})$ into a Fourier series,

$$\chi(\boldsymbol{r}) = \sum_{\boldsymbol{h}} \chi_{\boldsymbol{h}} \mathrm{e}^{\mathrm{i}\boldsymbol{h}\cdot\boldsymbol{r}} . \tag{2.79}$$

The Fourier polarizability coefficient can then be expressed as

$$\chi_{\boldsymbol{h}} = -\frac{\lambda^2 r_0}{\pi V_{\mathrm{u}}} F_{\boldsymbol{h}} , \tag{2.80}$$

where V_{u} is the volume of the elementary unit cell.

Since $\chi(\boldsymbol{r})$ shows the full symmetry of the crystal lattice – e.g. translational symmetry – the eigensolutions of the stationary wave equation (2.76) also exhibit translational symmetry. This leads to the concept of Bloch waves,[8] and the most general solution can be written as a Fourier series (plane waves)

$$\boldsymbol{E}(\boldsymbol{r}) = \mathrm{e}^{\mathrm{i}\boldsymbol{K}_0\cdot\boldsymbol{r}} \sum_{\boldsymbol{h}} \boldsymbol{E}_{\boldsymbol{h}} \mathrm{e}^{\mathrm{i}\boldsymbol{h}\cdot\boldsymbol{r}} = \mathrm{e}^{\mathrm{i}\boldsymbol{K}_0\cdot\boldsymbol{r}} \boldsymbol{u}_{\boldsymbol{K}_0}(\boldsymbol{r}) . \tag{2.81}$$

The Bloch function $\boldsymbol{u}_{\boldsymbol{K}_0}(\boldsymbol{r})$ exhibits the translational symmetry of the crystal lattice. In order to make clear the difference between the crystal and the vacuum, capital letters are used here for the wave vectors inside the crystal, whereas lower-case letters are used for wave vectors outside the crystal. By using

[7] With this approximation, we neglect qualitative differences between $\boldsymbol{E}(\boldsymbol{r})$ and $\boldsymbol{D}(\boldsymbol{r})$. Usually, different wave equations are formulated for $\boldsymbol{E}(\boldsymbol{r})$ and $\boldsymbol{D}(\boldsymbol{r})$. The error is of the order of $\chi(\boldsymbol{r}) \approx 10^{-5}$. It is also noteworthy that if we neglect the vector character of $\boldsymbol{E}(\boldsymbol{r})$, the wave equation (2.76) is formally equivalent to Schrödinger's equation. The latter describes dynamical diffraction of electrons, for example.

[8] The name "Bloch waves" goes back to Felix Bloch, who in 1928 in his doctoral thesis used this concept to solve Schrödinger's equation for electrons in a crystal lattice. However, P.P. Ewald had already used the same ansatz for his dynamical theory of X-ray diffraction.

the ansatz (2.81) and the Fourier series (2.79), the wave equation (2.76) can be transformed into an infinite set of linear algebraic equations,

$$\frac{\boldsymbol{K_h}^2 \boldsymbol{E}_{[\boldsymbol{h},\boldsymbol{h}]} - k^2 \boldsymbol{E_h}}{k^2} = \sum_{\boldsymbol{h}'} \chi_{\boldsymbol{h}-\boldsymbol{h}'} \boldsymbol{E}_{\boldsymbol{h}'} \, , \tag{2.82}$$

where $\boldsymbol{E}_{[\boldsymbol{h},\boldsymbol{h}]}$ is the component of $\boldsymbol{E_h}$ perpendicular to $\boldsymbol{K_h} \equiv \boldsymbol{K}_0 + \boldsymbol{h}$.

Equations (2.82) and (2.81) are the basic equations of the dynamical theory of X-ray diffraction, and they are generally valid. However, they cannot be solved exactly, and the Fourier expansion of the wave field $\boldsymbol{E}(\boldsymbol{r})$ has to be truncated after a finite number of terms. If only the zeroth term of the expansion (2.81) (i.e. $\boldsymbol{h} = 0$) is considered, only one (plane) wave is present inside the crystal. This solution is, therefore, often referred to as the *one-beam case*. The interaction of this wave with the crystal is mediated by the zeroth Fourier coefficient χ_0, and the crystalline structure is not "visible". The solutions obtained – the incoming, reflected, and transmitted waves – have already been described in Sect. 2.2.2. In that case the (coherent) reflectivity r_f and transmittivity t_f can be then calculated by using the Fresnel coefficients given in (2.17) and (2.25), and Parratt's recursion formalism [32] can be used for a multilayered sample.

The one-beam approximation is valid if there is no diffraction inside the crystal. If the sample fulfills the Bragg condition at the reciprocal-lattice point $\boldsymbol{h} \neq 0$, the diffracted wave $\boldsymbol{E_h}\mathrm{e}^{\mathrm{i}\boldsymbol{h}\cdot\boldsymbol{r}}$ has to be taken into account. In this case the amplitudes of the transmitted wave $\boldsymbol{E}_0$ and of the diffracted wave $\boldsymbol{E_h}$ are much stronger than all other amplitudes, which can be neglected.[9] The Fourier expansion (2.81) can be then reduced to

$$\boldsymbol{E}(\boldsymbol{r}) = \mathrm{e}^{\mathrm{i}\boldsymbol{K}_0\cdot\boldsymbol{r}} \left(\boldsymbol{E}_0 + \boldsymbol{E_h}\mathrm{e}^{\mathrm{i}\boldsymbol{h}\cdot\boldsymbol{r}} \right) = \boldsymbol{E}_0 \mathrm{e}^{\mathrm{i}\boldsymbol{K}_0\cdot\boldsymbol{r}} + \boldsymbol{E_h}\mathrm{e}^{\mathrm{i}\boldsymbol{K_h}\cdot\boldsymbol{r}} \, . \tag{2.83}$$

By putting the two waves with amplitudes $\boldsymbol{E}_0$ and $\boldsymbol{E_h}$ into (2.82), two coupled linear homogeneous equations of the form

$$[k^2(1+\chi_0) - K_0^2]\boldsymbol{E}_0 + k^2 P \chi_{-\boldsymbol{h}} \boldsymbol{E_h} = 0 \, , \tag{2.84a}$$

$$k^2 P \chi_{\boldsymbol{h}} \boldsymbol{E}_0 + [k^2(1+\chi_0) - K_{\boldsymbol{h}}^2]\boldsymbol{E_h} = 0 \, . \tag{2.84b}$$

are obtained, where P is the polarization factor given in (2.29). As can be seen, the coupling of the wave fields with the crystal is now mediated by χ_0, $\chi_{-\boldsymbol{h}}$, and $\chi_{\boldsymbol{h}}$. The appearance of both $\chi_{\boldsymbol{h}}$ and $\chi_{-\boldsymbol{h}}$ accounts for diffraction processes with diffraction vectors $\boldsymbol{h}$ and $-\boldsymbol{h}$.

---

[9] In principle, it is possible that three strong waves are present inside the crystal. The third wave $\boldsymbol{E}_{\boldsymbol{h}'}$ can be excited by rotating the sample around the reciprocal-lattice vector $\boldsymbol{h}$. The Bragg equation can then be simultaneously fulfilled at two different reciprocal-lattice points $\boldsymbol{h}$ and $\boldsymbol{h}'$. Renninger [90] was the first to demonstrate this case experimentally and discovered the *Umweganregung*. The *three-beam case* is a special case of dynamical diffraction and can be used to determine the phase of the diffracted wave [91, 92].

Nontrivial solutions for the amplitudes $\boldsymbol{E}_0$ and $\boldsymbol{E}_{\boldsymbol{h}}$ exist only if the determinant of (2.84) vanishes:

$$[k^2(1+\chi_0) - K_0^2][k^2(1+\chi_0) - K_{\boldsymbol{h}}^2] - k^4 P^2 \chi_{\boldsymbol{h}} \chi_{-\boldsymbol{h}} \,. \tag{2.85}$$

For the calculation of the electron band structure of a crystal, it is common to express the solutions in the form $E = E(\boldsymbol{K})$, i.e. energy vs. momentum. In a diffraction experiment, mostly monochromatic X-rays (or electrons) are used, and the intensity of diffracted waves is probed as a function of the angle of incidence. The dynamical theory of X-ray diffraction expresses the manifoldness of the solutions of (2.85) by the *dispersion surface*, which describes iso-energy surfaces in reciprocal space. We shall not go into any more details, since extensive discussions can be found in the literature [2, 93, 94].

It is important to mention that the solutions for the wave field $\boldsymbol{E}(\boldsymbol{r})$ depend on the boundary conditions for the wave fields and wave vectors at interfaces and surfaces. The boundary conditions couple the waves inside and outside the crystal. Therefore, the solutions in vacuum depend on the scattering geometry. The dynamical equations can be solved not only for a perfect semi-infinite crystal as sketched above, but also for multilayers. In this case the boundary conditions at each interface can be expressed as a matrix product. By using this matrix formalism, dynamical theory can be applied to laterally homogeneous ($\varrho(\boldsymbol{r}) = \varrho(z)$) layer systems [8, 10, 11, 12, 38]. In that case the wave field depends solely on the vertical z coordinate (growth direction). The matrix formalism is similar to the recursion method of Parratt [32]; however, four wave fields have to be considered. In the case of coplanar diffraction (two-beam case), Tagaki [95] and Taupin [96] have developed a theory that allows an approximate treatment of slightly strained layer systems. Also, (multilayer) surface gratings have been fully solved by dynamical theory [97, 98, 99].

We shall not proceed further with details related to dynamical theory. Excellent descriptions can be found in several textbooks, e.g. [2, 13, 93, 100]. For the reflection scattering geometry, the main consequences of dynamical theory can be summarized as:

- The amplitude of the incident wave decreases with depth below the sample surface (extinction).
- The diffracted intensity builds up from the interior of the sample towards the surface.
- In the case of grazing incidence/exit, the wave vectors outside and inside the crystal are significantly different in their directions (refraction).
- Below a critical angle, total external reflection occurs.

2.5 The Distorted-Wave Born Approximation

Unfortunately, the wave equation (2.76) cannot be solved exactly in most experimentally relevant cases. However, as will be shown in this section, it can be solved iteratively. For that purpose, the scattering potential $\hat{\mathbf{V}}(\boldsymbol{r})$ defined in (2.75) is divided into two parts:[10]

$$\hat{\mathbf{V}} = \hat{\mathbf{V}}_{\mathrm{A}} + \hat{\mathbf{V}}_{\mathrm{B}} \,. \tag{2.86}$$

The first term $\hat{\mathbf{V}}_{\mathrm{A}}$ describes the scattering by the perfect, undisturbed part of the sample. $\hat{\mathbf{V}}_{\mathrm{A}}$ can be chosen arbitrarily, as long as the equation

$$(\Delta + k^2)\boldsymbol{E}^{(\mathrm{A})}(\boldsymbol{r}) = \hat{\mathbf{V}}_{\mathrm{A}}(\boldsymbol{r})\boldsymbol{E}^{(\mathrm{A})}(\boldsymbol{r}) \tag{2.87}$$

can be solved exactly in the framework of dynamical theory. The second term, $\hat{\mathbf{V}}_{\mathrm{B}}$, can then be regarded as a (small) perturbation of the "perfect" scattering potential $\hat{\mathbf{V}}_{\mathrm{A}}$. We can express this in terms of the dielectric polarizability:

$$\hat{\mathbf{V}} = -k^2(\chi_{\mathrm{A}}(\boldsymbol{r}) + \delta\chi_{\mathrm{B}}(\boldsymbol{r})) \,, \tag{2.88}$$

where χ_{A} is the dielectric polarizability of the ideal undisturbed system and $\delta\chi_{\mathrm{B}}$ is the disturbance of it. Let us denote by $E_{\mathrm{s}}^{(-\mathrm{A})}(\boldsymbol{r})$ and $E_0^{(\mathrm{A})}(\boldsymbol{r})$ two independent solutions of (2.87). $E_0^{(\mathrm{A})}(\boldsymbol{r})$ describes the wave inside the sample that is excited by the incident plane wave $E_0(\boldsymbol{r}) = \mathrm{e}^{\mathrm{i}\boldsymbol{k}\cdot\boldsymbol{r}}$, whereas $E_{\mathrm{s}}^{(-\mathrm{A})}(\boldsymbol{r})$ is created by the vacuum wave $E_{\mathrm{s}}^{(-)}(\boldsymbol{r})$ that corresponds to the scattered wave.[11] The differential scattering cross section can then be expressed as [13]

$$\frac{\mathrm{d}\sigma}{\mathrm{d}\Omega}(\boldsymbol{q}) = \frac{1}{16\pi^2}\left|\langle E_{\mathrm{s}}^{(-\mathrm{A})}|\hat{\mathbf{V}}_{\mathrm{A}}|E_0\rangle + \langle E_{\mathrm{s}}^{(-\mathrm{A})}|\hat{\mathbf{V}}_{\mathrm{B}}|E_0^{(\mathrm{A})}\rangle\right|^2 \,. \tag{2.89}$$

The first term on the right-hand side represents coherent scattering by the perfect part (undisturbed potential $\hat{\mathbf{V}}_{\mathrm{A}}$) of the system. In the case of a perfect substrate this coherent signal is a δ-function-like expression, whereas a multilayer gives a rodlike expression which extends along the interface normal and is δ-function-like perpendicular to it.

The second term of (2.89) describes diffuse scattering from the disturbed potential $\hat{\mathbf{V}}_{\mathrm{B}}$. Here, the solution of the wave equation inside the crystal, $E_0^{(\mathrm{A})}(\boldsymbol{r})$, is scattered into $E_{\mathrm{s}}^{(-\mathrm{A})}(\boldsymbol{r})$. The second term thus describes an interaction between two exact solutions of the undisturbed case, and the coupling is mediated by the disturbed scattering potential $\hat{\mathbf{V}}_{\mathrm{B}}$. Equation (2.89) implies that the diffuse scattering from the disturbed regions and the coherent scattering from the perfect part of the sample can be treated independently. This is very advantageous for calculating the diffuse intensity.

[10] The nomenclature given in the textbook by Holý et al. [13] is used here.

[11] Note that the reciprocity principle of optics [101] enters into the equations, i.e. the scattering cross section is invariant under time inversion (indicated by the index "−"). Time inversion means that the scattered vacuum wave $E_{\mathrm{s}}^{(-)}(\boldsymbol{r})$ illuminates the sample in the reverse direction and creates the wave field $E_{\mathrm{s}}^{(-\mathrm{A})}(\boldsymbol{r})$ inside the sample.

The DWBA can be described in words as follows. The incident vacuum plane wave produces a wave field inside the sample. This wave field is treated as being the exact solution for the undisturbed sample. It is thus created by dynamical scattering from the undisturbed potential. In the next step this wave field is (diffusely) scattered by the disturbance inside the sample. After the scattering process, the resulting scattered waves leave the crystal (i.e. they are again dynamically scattered by the undisturbed potential).

The DWBA represents first-order perturbation theory. The exact solutions of (2.87) are calculated dynamically, whereas the diffuse scattering from the disturbed potential $\hat{\mathbf{V}}_{\mathrm{B}}$ is simply calculated kinematically. The DWBA approach is sketched schematically in Fig. 2.12.

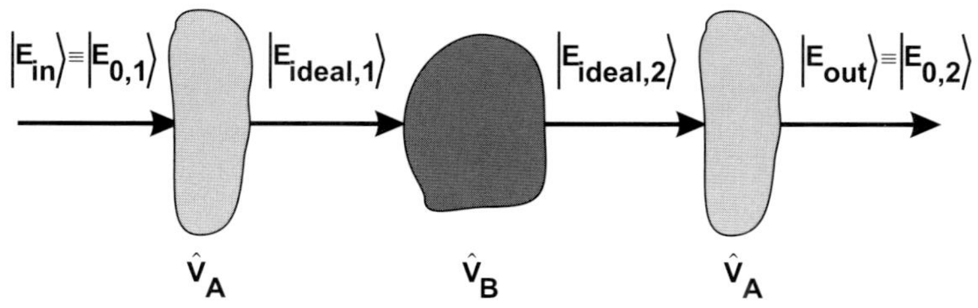

Fig. 2.12. Schematic illustration of the scattering process as approximated by the DWBA (after [13])

If we replace the wave fields $E_0^{(-\mathrm{A})}(\boldsymbol{r})$ and $E_{\mathrm{s}}^{(-\mathrm{A})}(\boldsymbol{r})$ inside the crystal by the vacuum plane waves $E_0(\boldsymbol{r}) = \mathrm{e}^{\mathrm{i}\boldsymbol{k}^{\mathrm{i}}\cdot\boldsymbol{r}}$ and $E_{\mathrm{s}}^{(-)}(\boldsymbol{r}) = \mathrm{e}^{-\mathrm{i}\boldsymbol{k}^{\mathrm{f}}\cdot\boldsymbol{r}}$, we obtain the first Born approximation. In that case the scattering cross section is identical to the kinematical approximation given in (2.33):

$$\frac{\mathrm{d}\sigma}{\mathrm{d}\Omega}(\boldsymbol{q}) = \frac{1}{16\pi^2}\left|\langle E_{\mathrm{s}}^{(-)}|\hat{\mathbf{V}}|E_0\rangle\right|^2 = r_0^2\left|\int \varrho(\boldsymbol{r})\mathrm{e}^{\mathrm{i}\boldsymbol{q}\cdot\boldsymbol{r}}\mathrm{d}V\right|^2 . \tag{2.90}$$

The polarization factor is missing in (2.90), since the term "grad div" has been omitted in the wave equation (2.74).

Generally, the solution of the Helmholtz equation can be written as a perturbation series (Born development scheme)

$$\boldsymbol{E}(\boldsymbol{r}) = \boldsymbol{E}^{(0)}(\boldsymbol{r}) + \boldsymbol{E}^{(1)}(\boldsymbol{r}) + \boldsymbol{E}^{(2)}(\boldsymbol{r}) + \dots , \tag{2.91}$$

where $\boldsymbol{E}^{(0)}(\boldsymbol{r})$ is the solution for the undisturbed case. The higher-order terms can be iteratively calculated from

$$\boldsymbol{E}^{(n)}(\boldsymbol{r}) = \int G(\boldsymbol{r}, \boldsymbol{r}')\hat{\mathbf{V}}_{\mathrm{B}}(\boldsymbol{r}')\boldsymbol{E}^{(n-1)}(\boldsymbol{r}')\,\mathrm{d}V' . \tag{2.92}$$

$G(\boldsymbol{r}, \boldsymbol{r}')$ is the Green's function of the homogeneous Helmholtz equation. Using the Born development scheme in (2.91), the DWBA can be calculated to higher orders. In general, the scattering is weak enough that only the first-order DWBA need be used. This is equivalent to the approximation that only single-scattering processes have to be taken into account. De Boer [47] has shown that the second-order DWBA leads to a general expression for the reflectivity and transmittivity for rough interfaces.

The exact solution of the wave equation is given by a set of different independent solutions (eigenstates). For a semi-infinite crystal, these solutions consist of two waves inside the crystal and three outside:

1. the incident vacuum plane wave,
2. the specularly reflected vacuum plane wave,
3. the diffracted vacuum plane wave,
4. the transmitted wave inside the crystal, and
5. the diffracted wave inside the crystal.

Thus, the matrix element as expressed in (2.89) represents a sum over all possible channels of the scattering process. For practical reasons, often not all of these channels are taken into account, since many of them are weak and can be neglected. The number of scattering channels that have to be considered depends on the sample structure and on the scattering geometry. For that reason, different levels of complexity of the DWBA have been reported in the literature depending on how many scattering channels are considered. This will be illustrated in the following examples.

2.5.1 Single-Channel DWBA

For grazing-incidence and/or grazing-exit geometries, significant effects due to refraction have to be considered. Therefore, in order to treat diffuse scattering from electron density fluctuations inside the substrate, the transmitted waves inside the substrate have to be considered. For grazing-incidence diffraction, this approach was initially developed by Vineyard [7] and was later completed by Dietrich and Wagner [102, 103], who obtained the most general result (valid for an arbitrary scattering geometry),

$$\frac{\mathrm{d}\sigma}{\mathrm{d}\Omega}(\boldsymbol{q}) = r_0^2\,|t_\mathrm{i}t_\mathrm{f}|^2 \left|\int \Delta\varrho(\boldsymbol{r})\mathrm{e}^{\mathrm{i}\boldsymbol{q}'\cdot\boldsymbol{r}}\,\mathrm{d}V\right|^2 , \tag{2.93}$$

where $\Delta\varrho(\boldsymbol{r})$ is the electron density of the disturbance and $\boldsymbol{q}' \equiv \boldsymbol{k}'_\mathrm{f} - \boldsymbol{k}'_\mathrm{i}$ denotes the scattering vector inside the crystal; t_i and t_f are the transmission functions given in (2.25) for the transmitted incident wave and the (time-inverted) transmitted outgoing wave. Since the transmittivity takes a value of 2 at the critical angle (when absorption and roughness are neglected), the signal can be enhanced by up to a factor of 16. Similarly, the enhancement of the diffuse intensity at the critical angle of total external reflection (for either the incident or the exit angle) leads to *Yoneda wings* [104] in the diffuse scattering.

As compared with the kinematical approach (2.33), all refraction effects at the surface are taken into account, while the scattering is still kinematical. This example is the simplest case of the DWBA, since a homogeneous medium is used as the reference "crystal" and only one scattering channel has been considered. Equation (2.93) can be interpreted as meaning that the incoming beam is refracted at the interface. The transmitted beam is then scattered from the electron density fluctuation $\Delta\varrho(\boldsymbol{r})$ and, after again being refracted at the surface, leaves the sample. This is illustrated schematically in Fig. 2.13.

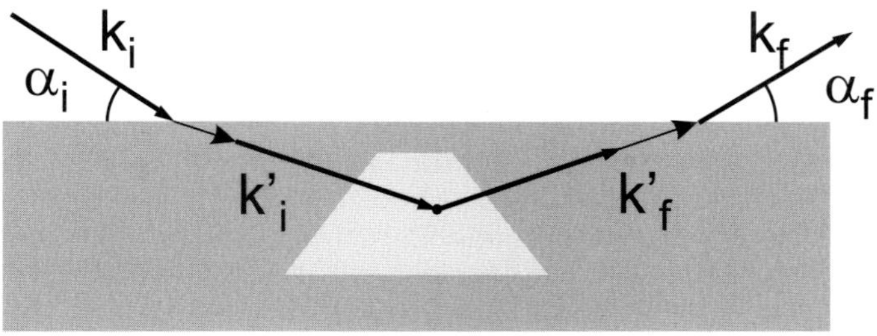

Fig. 2.13. Schematic illustration of single-channel DWBA for grazing-incidence small-angle X-ray scattering (GISAXS) from buried islands

Sinha et al. [9] have employed the DWBA to treat diffuse scattering originating from rough surfaces. Here, the perfect scattering potential is that of a laterally homogeneous semi-infinite medium with a flat surface, i.e. $\chi(\boldsymbol{r}) = \chi(z)$, whereas the disturbance is the electron density fluctuations (roughness) at the interfaces. In order to approach the experimental conditions as closely as possible, graded surfaces have often been used for the reference state. We shall discuss this case in more detail in Sect. 2.6. Sometimes even more complicated reference states have been used to describe scattering from gratings [105].

Stangl et al. [106] have used the same approach to describe the diffuse scattering from buried islands in the GISAXS geometry. For buried islands (embedded in a homogeneous medium), we can choose a laterally homogeneous medium, i.e. $\chi(\boldsymbol{r}) = \chi(z) = \chi_0$, as the perfect system, whereas the disturbance is given by local electron density fluctuations that are caused by the buried dots. Since the dots are buried, only the transmitted wave inside the medium has to be considered, and the wave reflected at the surface and underlying interfaces is neglected for the diffuse scattering. Consequently, the diffuse intensity in the GISAXS geometry can be expressed as [106, 107]

$$\frac{\mathrm{d}\sigma}{\mathrm{d}\Omega}(\boldsymbol{q}) = r_0^2\,\Delta\varrho^2 |t_\mathrm{i} t_\mathrm{f}|^2 |\Omega^\mathrm{FT}(\boldsymbol{q}')|^2 G(\boldsymbol{q})\;, \tag{2.94}$$

where $\Delta\varrho$ denotes the difference between the electron density of the buried structure and that of the surrounding material.

2.5.2 GISAXS from Free-Standing Islands: 5-Channel DWBA

The specular beam (at the surface and at interfaces) is missing in the kinematical theory and in the DWBA as presented in (2.93). However, this wave can also be scattered diffusely, leading to enhanced diffuse intensity in its vicinity. The inclusion of the specular beam in the framework of the DWBA was first performed by Kaganer et al. [108] for the evaluation of interface roughness. Here we shall focus on GISAXS from free-standing islands. In this case the diffuse scattering of the direct and specularly reflected beams is most important for scattering from the islands, whereas the transmitted beam has to be considered only for scattering inside the substrate.

Consequently, the following five channels, depicted in Fig. 2.14, have to be considered:

1. direct diffuse scattering from the island,
2. diffuse scattering from the island followed by specular reflection from the surface,
3. specular reflection from the surface followed by diffuse scattering from the island,
4. specular reflection from the surface followed by diffuse scattering from the island followed by specular reflection from the surface, and

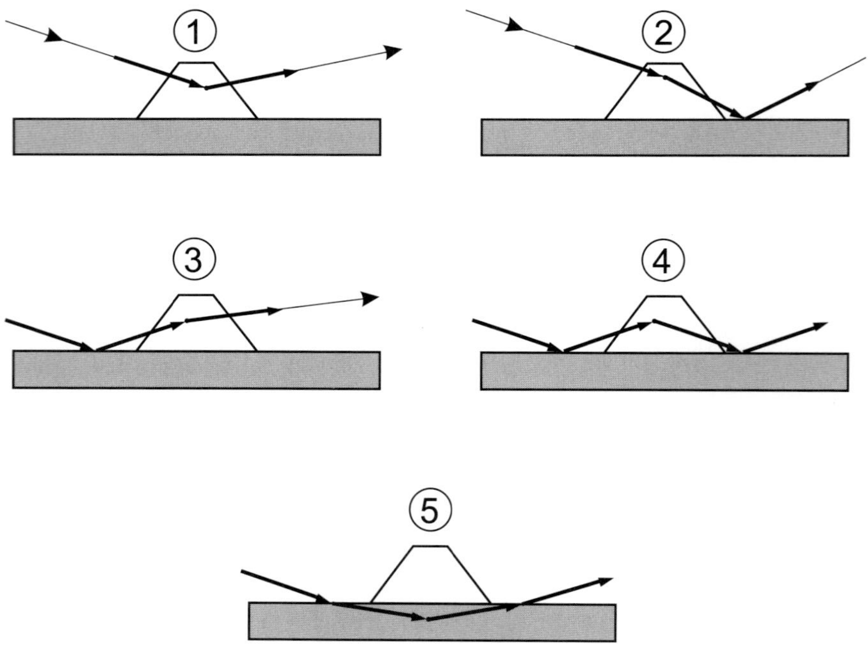

Fig. 2.14. Schematic view of the five channels for scattering from free-standing islands and the surrounding substrate

5. penetration into the substrate followed by diffuse scattering inside the substrate followed by transmission into the vacuum.

We can write the diffuse differential scattering cross section as

$$\frac{\mathrm{d}\sigma}{\mathrm{d}\Omega}(\boldsymbol{q}) = r_0^2 S(\boldsymbol{q}) \,, \tag{2.95}$$

where

$$S(\boldsymbol{q}) = S(\boldsymbol{q}_{\|}, k_z^{\mathrm{f}}, k_z^{\mathrm{i}}) = \left| A_1 + A_2 + A_3 + A_4 + A_5 \right|^2 . \tag{2.96}$$

The five amplitudes can be expressed as

$$A_1(\boldsymbol{q}) = \int\limits_{\text{island}} \varrho_{\text{dif}}(\boldsymbol{r}) \mathrm{e}^{\mathrm{i}(\boldsymbol{q}_{\|}+\boldsymbol{q}_1^{\perp})\cdot\boldsymbol{r}} \,\mathrm{d}V \,, \tag{2.97a}$$

$$A_2(\boldsymbol{q}_{\|}, k_z^{\mathrm{f}}) = r_{\mathrm{f}} \int\limits_{\text{island}} \varrho_{\text{dif}}(\boldsymbol{r}) \mathrm{e}^{\mathrm{i}(\boldsymbol{q}_{\|}+\boldsymbol{q}_2^{\perp})\cdot\boldsymbol{r}} \,\mathrm{d}V \,, \tag{2.97b}$$

$$A_3(\boldsymbol{q}_{\|}, k_z^{\mathrm{i}}) = r_{\mathrm{i}} \int\limits_{\text{island}} \varrho_{\text{dif}}(\boldsymbol{r}) \mathrm{e}^{\mathrm{i}(\boldsymbol{q}_{\|}+\boldsymbol{q}_3^{\perp})\cdot\boldsymbol{r}} \,\mathrm{d}V \,, \tag{2.97c}$$

$$A_4(\boldsymbol{q}_{\|}, k_z^{\mathrm{i}}, k_z^{\mathrm{f}}) = r_{\mathrm{i}} r_{\mathrm{f}} \int\limits_{\text{island}} \varrho_{\text{dif}}(\boldsymbol{r}) \mathrm{e}^{\mathrm{i}(\boldsymbol{q}_{\|}+\boldsymbol{q}_4^{\perp})\cdot\boldsymbol{r}} \,\mathrm{d}V \,, \tag{2.97d}$$

$$A_5(\boldsymbol{q}_{\|}, k_z^{\mathrm{i}}, k_z^{\mathrm{f}}) = t_{\mathrm{i}} t_{\mathrm{f}} \int\limits_{\text{substrate}} \varrho_{\text{dif}}(\boldsymbol{r}) \mathrm{e}^{\mathrm{i}(\boldsymbol{q}_{\|}+\boldsymbol{q}_5^{\perp})\cdot\boldsymbol{r}} \,\mathrm{d}V \,. \tag{2.97e}$$

The diffuse intensity given by (2.97) depends on the modified Fresnel reflection coefficients r_{i}, r_{f} and transmission coefficients t_{i}, t_{f}, where the subscripts i and f refer to the incoming and outgoing waves, respectively (see (2.25)). A very important point is that the vertical momentum transfers $\boldsymbol{q}^{\perp}$ are different for the five amplitudes. They are summarized in Table 2.2. Therefore, the diffuse intensity within the 5-channel DWBA approach is a function of both $\boldsymbol{q}$ and the angle of incidence α_{i}.

Table 2.2. Vertical momentum transfer $\boldsymbol{q}_j^{\perp}$ for the five different scattering amplitudes used in (2.97)

Channel	$\boldsymbol{q}_j^{\perp}$
$j = 1$	$+k_z^{\mathrm{f}} - k_z^{\mathrm{i}}$
$j = 2$	$-k_z^{\mathrm{f}} - k_z^{\mathrm{i}}$
$j = 3$	$+k_z^{\mathrm{f}} + k_z^{\mathrm{i}}$
$j = 4$	$-k_z^{\mathrm{f}} + k_z^{\mathrm{i}}$
$j = 5$	$+k'^{\mathrm{f}}_z - k'^{\mathrm{i}}_z$

The reflectivities $r_\mathrm{i}, r_\mathrm{f}$ and the transmittivities $t_\mathrm{i}, t_\mathrm{f}$ have the largest impact on the diffuse intensities when the angle of incidence α_i is chosen close to the critical angle of total external reflection α_c. At higher angles of incidence the reflectivities are small and can be neglected. In this case only the first and the fifth term on the right-hand side of (2.96) have to be considered, and the same result as predicted by the kinematical approach is obtained.

Rauscher et al. [109] have derived a similar expression. They were treating GISAXS from free-standing islands grown on a homogeneous (when one averages over the atomic structure) substrate and neglected the influence of the substrate. In this case only the direct and reflected beams need to be considered and A_5 vanishes. In addition, $\varrho_\mathrm{dif}(\boldsymbol{r})$ is replaced by ϱ_0 (the constant electron density inside the islands) and the differential scattering cross section can be written as

$$\left(\frac{\mathrm{d}\sigma}{\mathrm{d}\Omega}\right)_\mathrm{GISAXS} = r_0^2 \varrho_0^2 \left| \Omega^\mathrm{FT}(\boldsymbol{q}_\|, q_1^\perp) + r_\mathrm{f} \Omega^\mathrm{FT}(\boldsymbol{q}_\|, q_2^\perp) \right.$$
$$\left. + r_\mathrm{i} \Omega^\mathrm{FT}(\boldsymbol{q}_\|, q_3^\perp) + r_\mathrm{i} r_\mathrm{f} \Omega^\mathrm{FT}(\boldsymbol{q}_\|, q_4^\perp) \right|^2 . \quad (2.98)$$

Note that the amplitudes of the five different scattering channels are summed coherently and may thus interfere with each other. They are simultaneously present for a single island. Spatial correlation within an ensemble of monodisperse islands can be taken into account by replacing (2.95) with

$$\frac{\mathrm{d}\sigma}{\mathrm{d}\Omega}(\boldsymbol{q}) = r_0^2 S(\boldsymbol{q}) G(\boldsymbol{q}) , \quad (2.99)$$

where $G(\boldsymbol{q})$ is the interference function defined in (2.53).

2.5.3 Grazing-Incidence Diffraction from Free-Standing Islands: 9- and 16-Channel DWBA

When wide-angle diffraction is to be treated, the DWBA approach described above for GISAXS has to be modified. Close to a reciprocal lattice point $\boldsymbol{h}$, the scattered amplitude is not mediated exclusively by χ_0; it is influenced mainly by $\chi_{\boldsymbol{h}}$. In the case of grazing-incidence diffraction (GID), the solution of the dynamical problem supplies us with three plane waves in vacuum. These are the incoming wave, specularly reflected wave, and diffracted wave, with amplitudes E_n^in and wave vectors $\boldsymbol{k}_n^\mathrm{in}$. For these incoming ("in") wave fields, inverted outgoing ("out") wave fields can be constructed, with amplitudes $\mathrm{E}_n^\mathrm{out}$ and wave vectors $\boldsymbol{k}_n^\mathrm{out}$ (Fig. 2.15). The wave vectors are related to the vacuum momentum transfer through

$$\boldsymbol{q} = -(\boldsymbol{k}_0^\mathrm{in} + \boldsymbol{k}_0^\mathrm{out}) . \quad (2.100)$$

The index n labels the incident wave ($n = 0$), the reflected wave ($n = 1$), and the diffracted wave ($n = 2$), respectively. The reader should bear in mind that these waves are the solutions for the perfect, undisturbed crystal.

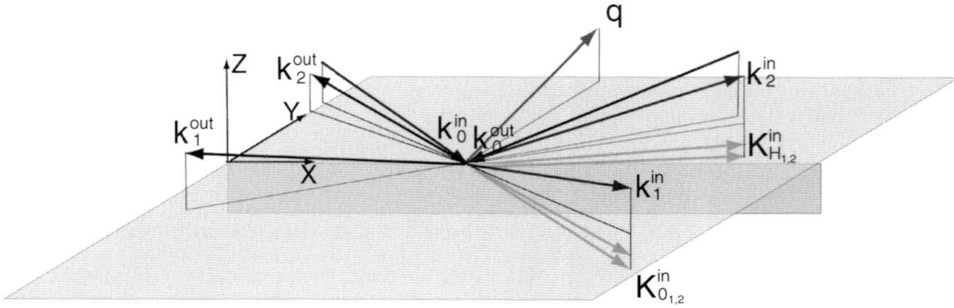

Fig. 2.15. Definition of wave vectors for incoming (in) and outgoing (out) waves for grazing-incidence diffraction. The indices 0, 1, and 2 correspond to incoming, reflected, and diffracted waves, respectively

According to dynamical theory, each of the incident plane waves, with wave vectors $\boldsymbol{k}_0^{\mathrm{in}}$ and $\boldsymbol{k}_0^{\mathrm{out}}$, creates a corresponding wave field inside the crystal, which consists of a pair of coupled plane waves with wave vectors $\boldsymbol{K}_{0(1,2)}^{\mathrm{in,out}}$ and $\boldsymbol{K}_{h(1,2)}^{\mathrm{in,out}}$ (Fig. 2.15) and corresponding amplitudes. These amplitudes can be calculated by solving the dynamical equations inside and outside the crystal. This consideration makes it clear that a calculation of the scattering from the semi-infinite substrate requires the consideration of four waves inside the crystal, and the DWBA coupling scheme provides us with 16 channels into which scattering can occur [38].

However, when the diffuse scattering from free-standing islands calculated, only three vacuum waves have to be taken into account: the incident, the specularly reflected, and the diffracted wave. In this case, the total amplitude can be expressed as [110]

$$A(\boldsymbol{q}) = \frac{k^2}{4\pi} \sum_{m=0}^{2} \sum_{n=0}^{2} \int E_m^{\mathrm{in}} E_n^{\mathrm{out}} \, \delta\chi_{mn}(\boldsymbol{r}) \mathrm{e}^{-\mathrm{i}(\boldsymbol{k}_m^{\mathrm{in}} + \boldsymbol{k}_n^{\mathrm{out}})\cdot \boldsymbol{r}} \, \mathrm{d}V \,, \tag{2.101}$$

where $\delta\chi_{mn}(\boldsymbol{r})$ is given by

$$\delta\chi_{mn}(\boldsymbol{r}) = \begin{cases} \delta\chi_0(\boldsymbol{r}) \text{ for } \dfrac{|\boldsymbol{k}_m^{\mathrm{in}} \cdot \boldsymbol{k}_n^{\mathrm{out}}|}{|\boldsymbol{k}_m^{\mathrm{in}}||\boldsymbol{k}_n^{\mathrm{out}}|} \approx 1 \\ \delta\chi_{\boldsymbol{h}}(\boldsymbol{r}) \text{ for } \dfrac{|\boldsymbol{k}_m^{\mathrm{in}} \cdot \boldsymbol{k}_n^{\mathrm{out}}|}{|\boldsymbol{k}_m^{\mathrm{in}}||\boldsymbol{k}_n^{\mathrm{out}}|} \approx \cos(2\theta_{\mathrm{B}}) \end{cases} , \tag{2.102}$$

(θ_{B} is the Bragg angle), and the differential scattering cross section is given by

$$\frac{\mathrm{d}\sigma}{\mathrm{d}\Omega}(\boldsymbol{q}) = r_0^2 |A(\boldsymbol{q})|^2 \,. \tag{2.103}$$

The double sum in (2.101) consists of nine terms that correspond to all possible types of coupling between $\boldsymbol{k}_m^{\mathrm{in}}$ and $\boldsymbol{k}_n^{\mathrm{out}}$. Therefore, we refer to this as the *9-channel* DWBA. In the case where the scattering signal of the island is separated (in reciprocal space) from the signal of the substrate, only four of the nine terms are essential: the incident and specularly reflected waves for each of $\boldsymbol{k}^{\mathrm{in}}$ and $\boldsymbol{k}^{\mathrm{out}}$. This case of four channels in GID can be treated similarly to GISAXS, as described in Sect. 2.5.2. The relevant channels are sketched in Fig. 2.14 (processes 1–4). In particular, the vertical momentum transfers $q_j^{\perp}$ given in Table 2.2 can be used for the various channels. We can then express the total amplitude as

$$A_1(\boldsymbol{q}) = \frac{k^2}{4\pi} \int\limits_{\mathrm{island}} \delta\chi_h(\boldsymbol{r}) \mathrm{e}^{\mathrm{i}(\boldsymbol{q}_{\parallel}+\boldsymbol{q}_1^{\perp})\cdot\boldsymbol{r}}\, \mathrm{d}V\,, \tag{2.104a}$$

$$A_2(\boldsymbol{q}_{\parallel}, k_z^{\mathrm{f}}) = r_{\mathrm{f}} \frac{k^2}{4\pi} \int\limits_{\mathrm{island}} \delta\chi_h(\boldsymbol{r}) \mathrm{e}^{\mathrm{i}(\boldsymbol{q}_{\parallel}+\boldsymbol{q}_2^{\perp})\cdot\boldsymbol{r}}\, \mathrm{d}V\,, \tag{2.104b}$$

$$A_3(\boldsymbol{q}_{\parallel}, k_z^{\mathrm{i}}) = r_{\mathrm{i}} \frac{k^2}{4\pi} \int\limits_{\mathrm{island}} \delta\chi_h(\boldsymbol{r}) \mathrm{e}^{\mathrm{i}(\boldsymbol{q}_{\parallel}+\boldsymbol{q}_3^{\perp})\cdot\boldsymbol{r}}\, \mathrm{d}V\,, \tag{2.104c}$$

$$A_4(\boldsymbol{q}_{\parallel}, k_z^{\mathrm{i}}, k_z^{\mathrm{f}}) = r_{\mathrm{i}} r_{\mathrm{f}} \frac{k^2}{4\pi} \int\limits_{\mathrm{island}} \delta\chi_h(\boldsymbol{r}) \mathrm{e}^{\mathrm{i}(\boldsymbol{q}_{\parallel}+\boldsymbol{q}_4^{\perp})\cdot\boldsymbol{r}}\, \mathrm{d}V\,, \tag{2.104d}$$

where the net vertical momentum transfers $\boldsymbol{q}^{\perp}$ are different for the four amplitudes and are summarized in Table 2.2.

As described in Sect. 2.3.3, the finite-element method can be used to calculate the strain field inside the islands and the substrate. Using (2.36), (2.39), and (2.43), the integrals in (2.101) and (2.104) are replaced by a discrete sum over all basic cells of the regular grid, and corresponding expressions similar to (2.44) are obtained.

Finally, we would like to emphasize again the most important result concerning scattering from free-standing islands grown on a substrate: scattering from the substrate has to be calculated by taking into account all four waves inside the crystal, whereas scattering from free-standing islands needs consideration of two waves, the incoming and specular waves, only. Consequently, scattering from the substrate includes 16 scattering channels, whereas scattering from free-standing islands requires four scattering channels.

2.6 Scattering from Surface and Interface Roughness

The term *roughness* means statistical fluctuations of the height profile of a surface or a buried interface. These changes in the height profile lead to fluctuations of the dielectric polarizability χ_0. As already pointed out in Sect. 2.2.2, roughness causes an exponential attenuation of the specularly reflected intensity by a factor $\mathrm{e}^{-4k_z k_z' \sigma^2}$, where σ is the rms roughness. However, the height profile of an interface is not only characterized by σ, but is also dependent on its spectrum of *spatial frequencies*. Since the specularly reflected beam has a vanishing lateral momentum transfer, it averages over all these frequencies across the interface (or surface). However, these frequencies are visible in the diffuse scattering. In other words, the characterization of the interface morphology regarding these spatial frequencies requires the investigation of diffuse scattering with a nonzero lateral momentum transfer.

Historically, the investigation of interface roughness was first carried out by using grazing-incidence scattering schemes close to the specular beam. The investigation of diffuse scattering close to the origin of reciprocal space is especially suited to that problem, since at small values of q the scattered intensities are sensitive to the interface morphology only, while strain effects can be neglected.

The inclusion of roughness in the theory of X-ray diffuse scattering was developed well before the first quantum dots or quantum wires were generated. At that time, the existence of roughness was an annoying phenomenon in that it might alter the electronic and optical properties of semiconductor thin films or cause severe limitations in the performance of multilayer X-ray mirrors with small periods. The common aim was therefore the suppression of roughness by choosing appropriate growth conditions.

The theoretical treatment of X-ray diffuse scattering from surface and interface roughness has to be carried out in the framework of the DWBA, which was introduced in Sect. 2.5. In principle, there are no reasons for treating the scattering from interface roughness separately from scattering from other mesoscopic structures. Nevertheless, we have devoted a separate section to this problem.

Arrays of quantum dots or quantum wires often exhibit a high degree of lateral ordering. By contrast, roughness is a phenomenon that implicitly includes a high degree of disorder. The reason for this is the large impact of growth kinetics on the evolution of roughness; that is, in its original sense, roughness is a state far from equilibrium. Several analytical models have been developed to describe the statistical electron density fluctuations. Among them are self-affine models with a cutoff length ξ which quantitatively describes the length up to which the self-affine regime is valid. This cutoff length is also referred to as the roughness correlation length and is different from the correlation lengths introduced in Sect. 2.3.4. For rather flat semiconductor surfaces, roughness is mostly connected with the step structure, and the evolution of steps is often related to the diffusion lengths on the surface and step–step interactions. However, even in the regime of very flat surfaces with a step structure, the self-affine models have proven to be a sufficient approximation to describe the roughness of surfaces and interfaces.

For systems with larger strains, well-ordered step bunches have been found, which evolve because of self-organizing growth processes that are driven by strain minimization. At this point of high ordering, the self-affine model will break down and the scattering has to be treated within the models described above. Roughness is therefore a very interesting phenomenon that is located between order and disorder.

This section aims to give an overview of the theoretical description of interface roughness and how it influences the diffuse scattering. The section is organized as follows. We start with some basic considerations on how roughness is incorporated into the DWBA approach. Then some general expressions for the differential scattering cross section are given. This is followed by a short treatise on how roughness can be described statistically. Then the influence of (inclined) vertical inheritance of interface roughness on the scattered intensity is discussed and compared with diffuse scattering originating from asymmetric vicinal interfaces. Finally, the effect of interfacial roughness on the diffuse scattering in a grazing-incidence diffraction geometry is discussed.

2.6.1 Scattering Cross Section

According to (2.88), the total dielectric polarizability is divided into a perfect part $\chi_{\text{flat}}(\boldsymbol{r})$ and small fluctuations $\delta\chi(\boldsymbol{r})$ related to the deviations from flat interfaces, i.e. interface roughness:

$$\chi(\boldsymbol{r}) = \chi_{\text{flat}}(\boldsymbol{r}) + \delta\chi(\boldsymbol{r})\,. \tag{2.105}$$

The intensity of diffuse scattering from roughness is then a statistical average over all fluctuations $\delta\chi(\boldsymbol{r})$ and can thus be expressed by a pair correlation function of the roughness of the type

$$C(\boldsymbol{r}_1, \boldsymbol{r}_2) = \langle \delta\chi(\boldsymbol{r}_1)\delta\chi(\boldsymbol{r}_2)\rangle\,. \tag{2.106}$$

At a rough interface, the electron density fluctuations are given by the height fluctuations at the interface. Therefore, the correlation function can be expressed in terms of the height profile of the interface. Later on, analytical models will be introduced that describe certain statistical properties of the height profile. However, before we proceed to the description of these correlation functions, expressions for the differential scattering cross section will be given.

In the DWBA approach, the differential scattering cross section for diffuse scattering (in the proximity of the specular beam) from a multilayer with N rough interfaces is given by [111]

$$\begin{aligned}\frac{\mathrm{d}\sigma}{\mathrm{d}\Omega}(\boldsymbol{q}) = \frac{Sk^4}{16\pi^2} \sum_{n,n'=1}^{N} \sum_{i,i'=\mathrm{r,t}} \sum_{j,j'=\mathrm{r,t}} \frac{C_{nij}C^*_{n'i'j'}}{q_{znij}q^*_{zn'i'j'}} \\ \times \int \mathrm{d}^2\boldsymbol{r}\left[\mathrm{e}^{q_{znij}q^*_{zn'i'j'}C_{nn'}(\boldsymbol{r})} - 1\right]\mathrm{e}^{-\mathrm{i}\boldsymbol{q}_\parallel\cdot\boldsymbol{r}}\,,\end{aligned} \tag{2.107}$$

where

$$C_{nij} = \delta\chi_0^n E_{ni}^{\text{in}} E_{nj}^{\text{out}} \mathrm{e}^{\mathrm{i}q_{znij}z_n - \sigma_n^2 q_{znij}^2/2} \, . \tag{2.108}$$

Here, $\delta\chi_0^n$ is the difference between the mean X-ray polarizabilities of the two layers forming the nth interface, and S is the area of the illuminated sample surface. The indices z and $\|$ denote the vector components perpendicular and parallel to the surface, respectively; $\boldsymbol{q}_{nij} = \boldsymbol{k}_{ni}^{\text{in}} + \boldsymbol{k}_{nj}^{\text{out}}$ is the momentum transfer on scattering. $\boldsymbol{k}_{ni}^{\text{in,out}}$ and $E_{ni}^{\text{in,out}}$ are the wave vectors and the amplitudes of the reflected (r) and transmitted (t) waves, respectively. These are the solutions of the specular-reflection problem in a flat multilayer for the real incident wave $\boldsymbol{K}_0$ (denoted by "in") and a virtual incident wave ("out"), which is the inverted scattered wave $\boldsymbol{k}_{\text{s}}$ (see Fig. 2.16). These "perfect" solutions correspond to flat interfaces and can be calculated by either the Parratt or the Abelès method [32, 34, 35]. The interface roughness is characterized by the rms roughness at each interface, $\sigma_n = \langle \delta z_n^2 \rangle^{1/2}$, and $C_{nn'}(\boldsymbol{r})$ is the height–height cross-correlation function of the interface relief. This will be discussed in detail in Sect. 2.6.2.

The differential scattering cross section given above describes the intensity distribution in all three dimensions q_x, q_y, and q_z in reciprocal space. In the experiment, a setup is often chosen where the intensity is integrated in the direction perpendicular to the plane of incidence. In this case (2.107) is integrated over the q_y component of the scattering vector, and we obtain

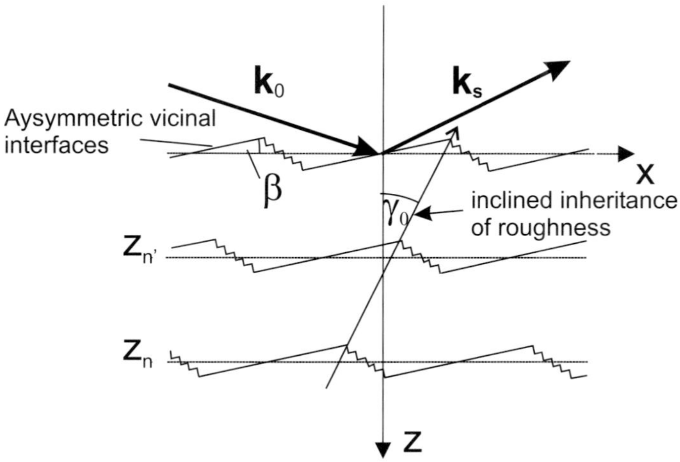

Fig. 2.16. Effects causing asymmetry in grazing-incidence X-ray scattering from multilayers: (i) steps in vicinal interfaces due to a small misorientation angle β between the interfaces and a crystallographic plane, and (ii) inclined inheritance of interfacial roughness with an angle γ_0. $\boldsymbol{k}_0$ and $\boldsymbol{k}_{\text{s}}$ are the wave vectors of the incident and scattered waves, respectively, x is the direction along the sample surface, and $\boldsymbol{z}$ is the normal vector of the mean surface. For clarity, the vertical scale is expanded relative to the horizontal scale

$$\frac{\mathrm{d}\sigma}{\mathrm{d}\Omega}(\boldsymbol{q}) = \frac{Sk^3}{8\pi} \sum_{n,n'=1}^{N} \sum_{i,i'=\mathrm{r,t}} \sum_{j,j'=\mathrm{r,t}} \frac{C_{nij} C^*_{n'i'j'}}{q_{znij} q^*_{zn'i'j'}}$$
$$\times \int_{-\infty}^{+\infty} \mathrm{d}x \left[\mathrm{e}^{q_{znij} q^*_{zn'i'j'} C_{nn'}(x)} - 1 \right] \mathrm{e}^{-\mathrm{i} q_x x} . \qquad (2.109)$$

Note that two waves have been used here – the reflected wave (r) and transmitted wave (t). The transmitted wave passing through interface n serves as the incoming wave illuminating interface $n+1$.

2.6.2 Statistical Description of Roughness: the Self-Affine Model

The statistical nature of the roughness of a static, laterally homogeneous interface can be described by a *height-difference* correlation function of the height profile $z(x, y)$, which is defined as

$$g(\boldsymbol{R}) \equiv \langle [z(x', y') - z(x, y)]^2 \rangle , \qquad (2.110)$$

where the average is performed over all pairs of points (x, y) and (x', y') on the interface that are separated by the distance vector $\boldsymbol{R}$ (see Fig. 2.17). The properties of $g(\boldsymbol{R})$ depend critically on the material system. Many isotropic solid interfaces can be specified by *self-affine* height profiles. In such a case the correlation function takes the form [112]

$$g(\boldsymbol{R}) = g(R) = AR^{2H} , \qquad (2.111)$$

where H denotes the *Hurst parameter* $(0 < H < 1)$ or roughness exponent. The features of the self-affine model are given by its scaling behavior. As can be seen, (2.111) is invariant with respect to the scaling transformation $R \rightarrow bR$ and $z \rightarrow zb^{-H}$. The self-affine scaling behavior can be justified by kinetic growth models (e.g. [113, 114]), and excellent articles [115, 116] reviewing these models can be found in the literature.

Usually, the self-affine description of the interface morphology is valid in only a restricted range of lateral length scales. On one hand, the self-affine

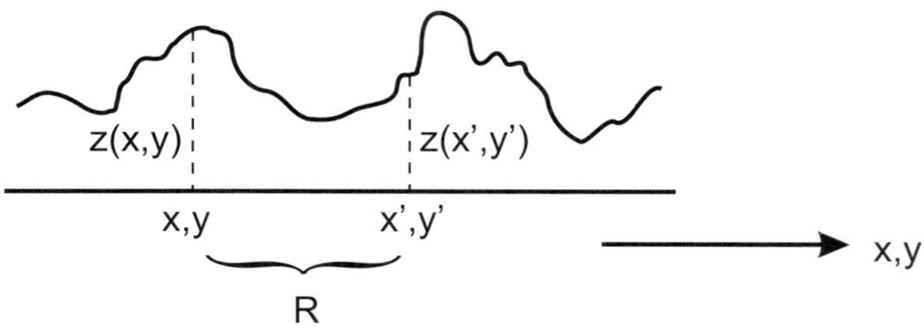

Fig. 2.17. Schematic view of a rough interface. The height profile can be described by a uniformly continuous function $z = z(x, y)$

picture will be certainly lost at the atomic length scale, where a discrete step structure may be present, for example. We shall discuss that case below in more detail. On the other hand, a finite value σ of the rms roughness has to be expected, since semiconductor samples exhibit a flat surface on a macroscopic length scale. In other words, an upper limit of $g(R)$ has to exist for $R \to \infty$. Therefore, the self-affine picture will also break down at macroscopic length scales. In order to fulfill this asymptotic requirement, a cutoff length ξ is introduced, and a possible height-difference correlation function is given by [9]

$$g(R) = 2\sigma^2\left[1 - \mathrm{e}^{-(R/\xi)^{2H}}\right] . \tag{2.112}$$

By using the relationship $g(R) = 2\langle z(x,y)^2\rangle - 2\langle z(x,y)z(x',y')\rangle = 2\sigma^2 - C(R)$, we have already exploited the fact that $\sigma^2 \equiv \langle z(x,y)^2\rangle$ is finite. $C(R)$ is the *height–height* correlation function,

$$C(R) = \sigma^2 \mathrm{e}^{-(R/\xi)^{2H}} . \tag{2.113}$$

Instead of describing roughness by correlation functions defined in real space, the equivalent concept of the spectral power density (see also (2.72)) is often used:

$$P^{\mathrm{affine}}(q_x, q_y) \equiv \left| \int z(x,y)\mathrm{e}^{\mathrm{i}(q_x x + q_y y)}\,\mathrm{d}x\,\mathrm{d}y \right|^2 = \int C(R)\mathrm{e}^{-\mathrm{i}\boldsymbol{q}_\parallel \cdot \boldsymbol{R}}\,\mathrm{d}^2 R , \tag{2.114}$$

which is defined in reciprocal space.

Note that the cutoff length ξ is usually referred to as the roughness *correlation length*, although it is of a different physical nature from the correlation length introduced in Sect. 2.3.4. The roughness correlation length quantifies the length scale up to which the morphology can be described by a self-affine model, whereas the correlation length defined in Sect. 2.3.4 quantifies the length scale up to which positional correlation is present. As a consequence, no lateral correlation peaks are observed in the diffuse scattering from self-affine roughness (with a cutoff). Instead, the presence of a continuous spectrum of spatial frequencies with a cutoff at low values ensures a rather smooth intensity profile of the diffuse scattering. However, the correlation length ξ determines the shape of the diffuse intensity. The smaller ξ is, the larger is the deviation of the diffuse scattering from the coherent truncation rod, and vice versa. This is visible in Fig. 2.18. Moreover, the diffuse scattering pattern is symmetrical with respect to the lateral scattering vector $q_\parallel$. This follows from the reciprocity theorem and the fact that self-affine roughness does not change when the surface is rotated azimuthally by 180°.

Generally, the expression *anisotropic* roughness (i.e. $C = C(\boldsymbol{R}) \neq C(R)$) is used for a system that exhibits different roughness parameters ξ and H

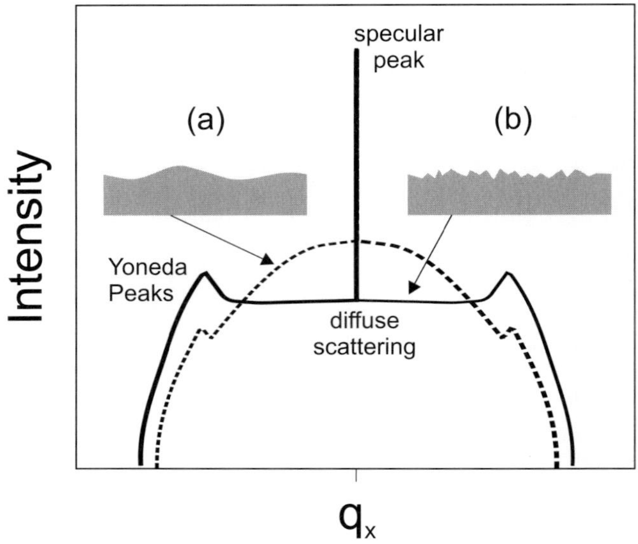

Fig. 2.18. Schematic view of diffuse scattering from a rough surface. At low spatial frequencies (**a**), i.e. high correlation lengths, it is concentrated into limited values of q_x, whereas at high spatial frequencies (**b**), the diffuse intensity is spread out widely along q_x

in different lateral directions across the interface. As we shall see later, this description is necessary for treating the roughness of vicinal interfaces.

2.6.3 Vertical Inheritance of Interface Roughness in Multilayers

So far, the statistical roughness properties of a single interface have been discussed. How can a multilayer with N interfaces be described? One possibility is to treat each interface as being independent of all other interfaces. In this case of *vertically uncorrelated* roughness, the number of free fitting parameters is very large: since each layer has to be described by three independent parameters (σ^2, ξ, and H), the total number of parameters that have to be experimentally determined is $3N$. As a consequence, a reliable data evaluation is practically impossible. Fortunately, the interface roughness of multilayers is often vertically correlated. This means that the interface morphology is – to some extent – vertically inherited during the growth procedure (Fig. 2.19).

The vertical correlation can be described by a *cross-correlation* function

$$C_{nn'}(\boldsymbol{R}) = \langle z_n(\boldsymbol{r}) z'_n(\boldsymbol{r}+\boldsymbol{R})\rangle \,, \qquad n \neq n'; \quad n, n' = 1, \ldots, N \,. \quad (2.115)$$

Different analytical expressions for $C_{nn'}(\boldsymbol{R})$ have been reported in the literature and have been extensively discussed [8, 108, 117, 118, 119, 120, 121, 122].

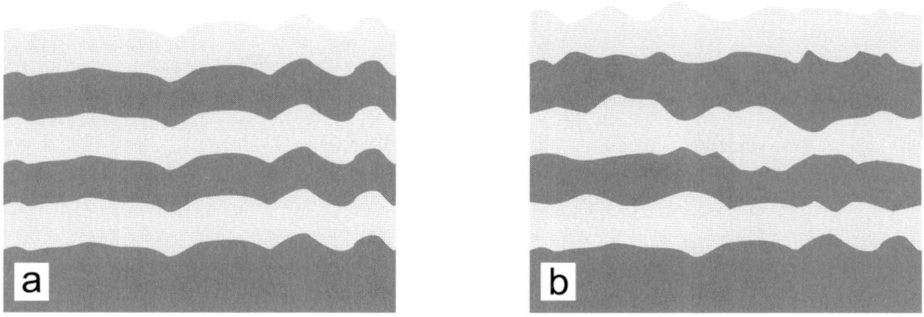

Fig. 2.19. Schematic representation of the interface morphology of a multilayer exhibiting (**a**) complete vertical correlation of roughness, and (**b**) vertically uncorrelated roughness

Some of these publications [108, 117, 118, 121, 122] use sophisticated models that allow selective vertical inheritance as a function of spatial frequency. The idea behind this is that small spatial frequencies are preferentially inherited, whereas large spatial frequencies are more or less random and thus uncorrelated. This behavior can be qualitatively justified by growth models (e.g. [113, 118]) and is mostly applied to rough amorphous multilayers exhibiting a large range of spatial frequencies. For rather flat semiconductor multilayers with high vertical correlation, the height–height cross-correlation function proposed by Ming et al. [120] has turned out to be a good description, and the total correlation function in one dimension can be written as

$$C_{nn'}(x) = \sigma^2 \exp[-(x/\xi)^{2H}] \exp(-|z_n - z_{n'}|/\xi_z) \,. \tag{2.116}$$

In this simplified model, each interface is treated as having identical lateral roughness parameters, whereas the vertical correlation of the interfaces n and n' depends simply on the distance between them $|z_n - z_n'|$, where z_n are the coordinates of the mean interfaces (Fig. 2.16). For $\xi_z = \infty$ the interface morphology is completely replicated, whereas for $\xi_z = 0$ the roughness is vertically uncorrelated.

It has been shown by Stearns [117] and Holý and Baumbach [8] that in the case of vertical correlation, the diffuse scattering from the interface roughness of a periodic superlattice is concentrated within *resonant diffuse scattering sheets* (RDS sheets, also called "bananas"), which are located in the vicinity of the coherent superlattice satellites. Similarly to diffuse scattering from a single interface, the vertical and lateral dimensions of the RDS sheets scale inversely with the vertical and lateral height–height correlation lengths, respectively. The effect of vertical correlation on the diffuse intensity pattern is shown in Fig. 2.20.

Equation (2.116) describes inheritance of roughness along the (mean) interface normal of the multilayer. As will be discussed later in Chap. 7, this exact vertical replication need not be fulfilled strictly. Instead, the interface

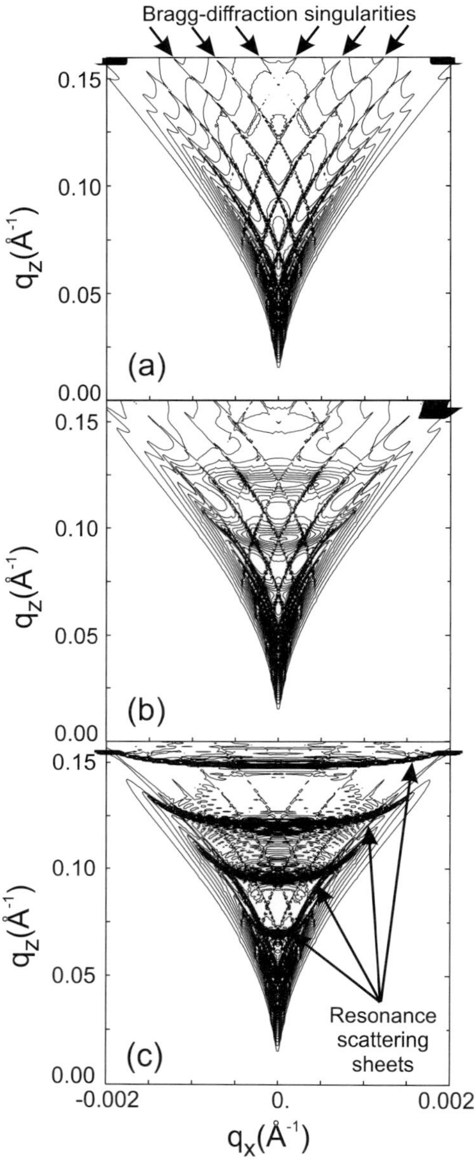

Fig. 2.20. Effect of vertical correlation of interfacial roughness in a superlattice. With increasing vertical correlation length, resonant diffuse scattering (RDS) sheets are formed. The calculations were performed for a 20 period AlAs/GaAs superlattice. (**a**) No vertical correlation, (**b**) vertical correlation of one superlattice period, (**c**) complete vertical correlation (after Kaganer et al. [108]). The intensity distribution is influenced by refraction effects, leading to a bowing of the RDS sheets in reciprocal space when the angle of incidence or exit is close to the critical angle of total external reflection

morphology can be inherited at a certain angle $\gamma_0 \neq 0$ which deviates from the exact vertical direction [111, 123, 124] (Fig. 2.16). If we neglect strain induced processes in strongly anisotropic systems [82], one reason for this behavior is the occurrence of a step-flow mode during epitaxial growth, which may lead to a net lateral transfer of steps in the step-flow direction. In order to take account of this phenomenon in the corresponding X-ray diffuse scattering, the interface–interface cross-correlation function is transformed as

$$C_{nn'}(x) \rightarrow C_{nn'}[x - \tan\gamma_0(z_n - z_{n'})] \, . \tag{2.117}$$

If we assume small roughness ($q_z\sigma \ll 1$), the integral in (2.109) can be replaced by the Fourier transform of the height–height cross-correlation function $C_{nn'}(x)$. Taking account of only the transmitted waves in the sum of (2.109) and using the expression (2.117), the following final formula is obtained:

$$\frac{\mathrm{d}\sigma}{\mathrm{d}\Omega}(\boldsymbol{q}) = \frac{Sk^3}{8\pi} \sum_{n,n'} C_{n\mathrm{tt}} C^*_{n'\mathrm{t}'\mathrm{t}'} \mathrm{e}^{-\mathrm{i}q_x \tan\beta(z_n - z_{n'})} P_{nn'}(q_x) \, , \tag{2.118}$$

where $P_{nn'}(q_x)$ is the one-dimensional Fourier transform of (2.117). Note that the consideration of inclined roughness transfer leads to an asymmetry of the corresponding spectral power density function $P_{nn'}(q_x)$ with respect to q_x. This is demonstrated in Fig. 2.21.

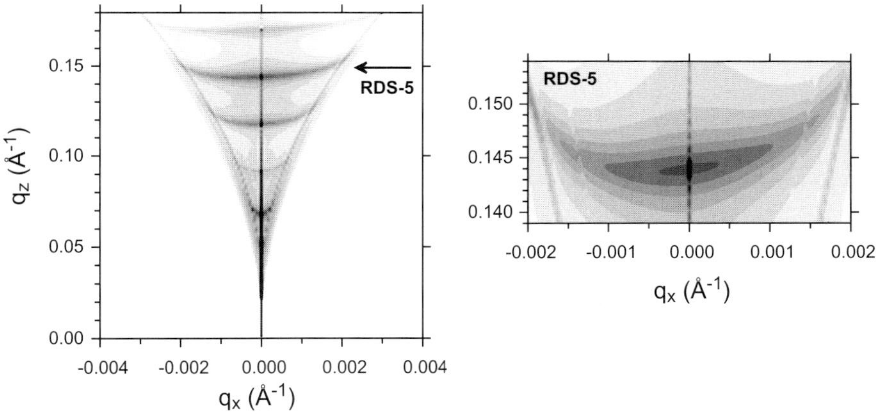

Fig. 2.21. Simulations of diffuse intensity in the model of inclined inheritance of roughness using the correlation function given in (2.117). The simulation was performed for an AlAs/GaAs superlattice with a surface miscut of 0.4°. The inclination angle was chosen as 45°. The picture has been taken from [36]

2.6.4 Statistical Description of Roughness: the Step Model

It has been suggested in the literature [40, 111, 119, 125, 126] that the self-affine model of roughness needs some modifications in the case of rather perfect interfaces with a step structure. For that purpose, Stettner et al. [40, 127, 128] have investigated the intermediate range between the self-affine case and the (hypothetical) case of equidistantly spaced steps. In the latter case, these authors found – as expected – sharp correlation peaks, but it was also discovered that even rather small fluctuations in the terrace widths lead to interfaces that can be well described by a self-affine height–height correlation function with a cutoff (Fig. 2.22).

Nevertheless, if the fluctuations of the terrace widths are not large enough, the validity of the self-affine model is lost. Considering a one-dimensional step distribution with mean terrace length ξ and step height d, Pukite et al. [129] have derived a general expression for the spectral power density function (in one dimension), which for small roughness can be written in the simple form [111]

$$P_{nn'}^{\mathrm{steps}}(q_x) = \frac{\xi q_z^2 d^2 \mathrm{e}^{-(|z_n - z_{n'}|/\xi_z)}}{(q_x\xi + q_z d)^2 + q_z^4 d^4/4} \, . \tag{2.119}$$

The most striking feature of this expression is its asymmetry with respect to q_x, i.e. $P_{nn'}(q_x) \neq P_{nn'}(-q_x)$. In particular $P_{nn'}(q_x)$ peaks at values $q_x = -\beta q_z$, where $\beta = d/\xi$ is the surface miscut of the sample (see Fig. 2.16). This means that there is enhanced intensity in the direction parallel to the terrace surface normal. This behavior has been theoretically predicted by Sinha et al. [126] and is consistent with experimental observations [130, 131,

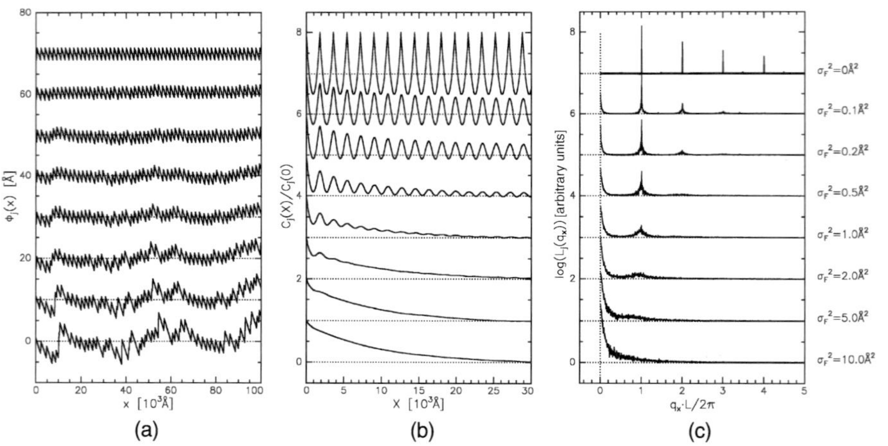

Fig. 2.22. (**a**) Simulated height profiles $z_j(x)$ for different terrace width distributions represented by the parameter σ_{f}, (**b**) corresponding normalized height–height correlation functions $C_j(X)/C_j(0)$, (**c**) the corresponding spectral power density functions $P_j(q_x)$. The figure has been taken from Stettner et al. [40, 128]

132]. The asymmetry of diffuse scattering may thus give some evidence of a stepped surface with a statistical terrace length distribution.[12] However, as shown in (2.118), inclined inheritance of roughness may also lead to a similar asymmetry.

If the terraces are not flat surfaces but undergo some local, random up and down fluctuations, the spectral power density has to be modified. In this case a self-affine fraction of roughness has to be added to the spectral power density:

$$P_{nn'}(q_x) = P_{nn'}^{\text{affine}}(q_x) + P_{nn'}^{\text{steps}}(q_x) \,. \tag{2.120}$$

Using this expression, Kondrashkina et al. [111] obtained

$$P_{nn'}(q_x) = \frac{2\xi q_z^2 \sigma_{\text{eff}}^2 \mathrm{e}^{-(|z_n - z_{n'}|/\xi_z)}}{(q_x\xi + q_z\beta\xi)^2 + q_z^4 d^4/4} \,, \tag{2.121}$$

where $\sigma_{\text{eff}}^2 = \sigma^2 + \beta^2\xi^2/2$.

When the small-roughness approximation ($q_z\sigma \ll 1$) is used, the integral in (2.109) can be replaced by the Fourier transform of the correlation function, i.e. the expression given in (2.121) can be used. If only the transmitted waves are considered and if refraction (which is present at angles larger than the critical angle) is neglected, we end up with the Born approximation:

$$\frac{\mathrm{d}\sigma}{\mathrm{d}\Omega}(\boldsymbol{q}) = \frac{Sk^3}{8\pi} \sum_{n,n'} \Delta\chi_0^n (\Delta\chi_0^n)^* \mathrm{e}^{-q_z^2(\sigma_n^2 + \sigma_n'^2)/2} \mathrm{e}^{-\mathrm{i}q_z(z_n - z_{n'})} P_{nn'}(q_x) \,. \tag{2.122}$$

Exemplary calculations of the diffuse scattering using (2.121) are shown in Fig. 2.23. The RDS sheets are shifted to the left side, causing an asymmetry with respect to q_x. These shifts can be clearly differentiated from the effect of inclined roughness inheritance, which, in contrast, leads to a shearing of the RDS sheets. This qualitative difference in the diffuse scattering enables one to clearly distinguish between inclined inheritance of roughness and the presence of asymmetric vicinal interfaces. Since both effects can in principle be present in multilayers grown on vicinal substrates, this distinction is essential.

2.6.5 Effect of Roughness on Grazing-Incidence Diffraction

So far, we have restricted ourselves to off-specular diffuse scattering (grazing-incidence X-ray scattering, GIXS). In principle, the same mathematical procedure can be applied to the diffuse scattering in grazing-incidence diffraction (GID). There are, however, two important differences in the treatment of diffuse scattering in GID and GIXS:

[12] This asymmetry in the diffuse scattering is equivalent to Friedel's law [133]. Since the morphology of a stepped surface is not symmetrical with respect to $x \to -x$, the corresponding diffuse intensity pattern is not symmetrical with respect to q_x.

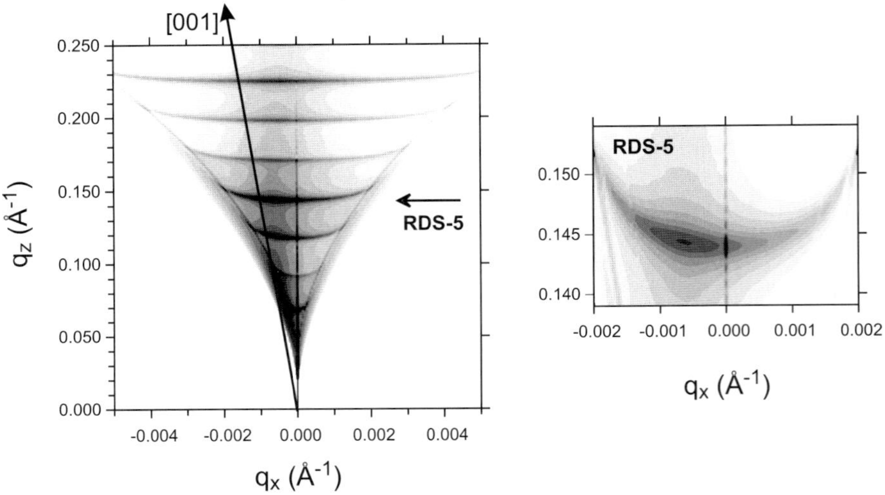

Fig. 2.23. Simulations of diffuse intensity for the step model using the correlation function given in (2.121). The simulation was performed for an AlAs/GaAs superlattice with a surface miscut of 0.4°. The picture has been taken from [36]

1. In the case of GIXS, the perfect solutions used in the DWBA are calculated by using the formalism of Parratt [32] or Abelès [34, 35] (see Sect. 2.2.2), while for GID the dynamical solutions for diffraction must be used. This enhances the number of waves that have to be taken into account from two to four, and the corresponding (4×4) matrices have to be used [38].
2. In the case of GIXS, an average over the atomistic structure of the sample is performed and only morphological fluctuations are visible. Therefore, the difference between the mean electron densities of the two materials at the interface, i.e. $\delta\chi(\boldsymbol{r}) = \delta\chi_0(\boldsymbol{r})$, is relevant. For GID, the fluctuation of the dielectric polarizability is not determined only by its zeroth Fourier expansion coefficient as given above, but it also depends on the difference between the Bragg scattering amplitudes [12]: $\delta\chi(\boldsymbol{r}) = \delta\chi_0(\boldsymbol{r}) + \delta\chi_h(\boldsymbol{r})\mathrm{e}^{\mathrm{i}\boldsymbol{h}\cdot\boldsymbol{r}} + \delta\chi_{-h}(\boldsymbol{r})\mathrm{e}^{-\mathrm{i}\boldsymbol{h}\cdot\boldsymbol{r}}$.

For more details the reader is referred to the article by Stepanov et al. [12].

References

1. P. P. Ewald, Ann. Phys. **49**, 1 (1916).
2. M. von Laue, *Röntgenstrahlinterferenzen* (Akademische Verlagsgesellschaft, Frankfurt am Main, 1960).
3. L. V. Azaroff, R. Kaplow, N. Kato, R. J. Weiss, A. J. C. Wilson, and R. A. Young, *X-Ray Diffraction* (McGraw-Hill, New York, 1974).

4. J. M. Cowley, *Diffraction Physics* (North-Holland, Amsterdam, 1975).
5. M. A. Krivoglaz, *Diffuse Scattering of X-Rays and Neutrons by Fluctuation* (Springer, Berlin, Heidelberg, 1996).
6. M. A. Krivoglaz, *X-Ray and Neutron Diffraction in Nonideal Crystals* (Springer, Berlin, Heidelberg, 1996).
7. G. H. Vineyard, Phys. Rev. B **26**, 4146 (1982).
8. V. Holý and T. Baumbach, Phys. Rev. B **49**, 10668 (1994).
9. S. K. Sinha, E. B. Sirota, S. Garoff, and H. B. Stanley, Phys. Rev. B **38**, 2297 (1988).
10. S. A. Stepanov, E. A. Kondrashkina, R. Köhler, D. V. Novikov, G. Materlik, and S. M. Durbin, Phys. Rev. B **57**, 4829 (1998).
11. S. A. Stepanov, U. Pietsch, and G. T. Baumbach, Z. Phys. B: Cond. Matter **96**, 341 (1995).
12. S. A. Stepanov, E. A. Kondrashkina, M. Schmidbauer, R. Köhler, J. U. Pfeiffer, T. Jach, and A. Y. Souvorov, Phys. Rev. B **54**, 8150 (1996).
13. V. Holý, U. Pietsch, and T. Baumbach, *High Resolution X-Ray Scattering from Thin Films and Multilayers*, vol. 149 of *Springer Tracts in Modern Physics* (Springer, Berlin, Heidelberg, 1999).
14. J. Als-Nielsen and D. McMorrow, *Elements of Modern X-Ray Physics* (Wiley, New York, 2001).
15. B. E. Warren, *X-Ray Diffraction* (Addison-Wesley, Reading, MA, 1969), reprinted by Dover, New York, 1990.
16. W. H. Zachariasen, *Theory of X-Ray Diffraction in Crystals* (Dover, New York, 1945).
17. R. Feidenhans'l, Surf. Sci. Rep. **10**, 105 (1989).
18. I. K. Robinson and D. J. Tweed, Rep. Prog. Phys. **55**, 599 (1992).
19. J. Daillant and M. Alba, Rep. Prog. Phys. **63**, 1725 (2000).
20. H. Hönl, Ann. d. Physik **18**, 538 (1933).
21. H. A. Kramers, Z. Phys. **53**, 422 (1929).
22. R. de L. Kronig, Z. Phys. **50**, 347 (1928).
23. R. de L. Kronig and H. A. Kramers, Z. Phys. **48**, 174 (1928).
24. G. Materlik, C. J. Sparks, and K. Fischer, eds., *Resonant Anomalous X-Ray Scattering* (North-Holland, Amsterdam, 1994).
25. Y. Waseda, *Anomalous X-Ray Scattering for Materials Characterization*, vol. 179 of Springer Tracts in Modern Physics (Springer, Berlin, Heidelberg, 2002).
26. A. H. Compton, Bull. Nat. Res. Counc. **20**, 48 (1922).
27. A. H. Compton, Phys. Rev. **20**, 84 (1922).
28. A. H. Compton, Phil. Mag. **45**, 1121 (1923).
29. H. Dosch, *Critical Phenomena at Surfaces and Interfaces*, vol. 126 of Springer Tracts in Modern Physics (Springer, Berlin, Heidelberg, 1992).
30. S. A. Stepanov, http://sergey.bio.aps.anl.gov/x0h.html.
31. M. Born and E. Wolf, *Principles of Optics* (Pergamon, Oxford, 1980).
32. L. G. Parratt, Phys. Rev. **95**, 359 (1954).
33. M. Tolan, *X-Ray Scattering from Soft-Matter Thin Films*, vol. 148 of Springer Tracts in Modern Physics (Springer, Berlin, Heidelberg, 1999).
34. F. Abelès, Ann. Phys. (Paris) **3**, 504 (1948).
35. F. Abelès, Ann. Phys. (Paris) **5**, 596 (1950).

36. R. Opitz (1999), *Charakterisierung der Grenzflächenmorphologie vizinaler AlAs/GaAs-Schichtsysteme mittels Röntgenstreuung*, Ph.D. thesis, Humboldt-Universität zu Berlin (Logos Verlag, Berlin, 1999).
37. F. Stanglmeier, B. Lengeler, W. Weber, H. Göbel, and M. Schuster, Acta Crystallogr. A **48**, 626 (1992).
38. S. A. Stepanov and R. Köhler, J. Appl. Phys. **76**, 7809 (1994).
39. J. Lekner, *Theory of Reflection* (Martinus Nijhoff, Dordrecht, 1987).
40. J. Stettner (1995), Ph.D. thesis, Universität Kiel (1995).
41. L. Névot and P. Croce, Phys. Rev. Appl. **15**, 761 (1980).
42. B. Vidal and P. Vincent, Appl. Opt. **23**, 1794 (1984).
43. R. Pynn, Phys. Rev. B **45**, 606 (1992).
44. P. Beckman and A. Spizzichino, *The Scattering of Electromagnetic Waves from Rough Surfaces* (Pergamon, Oxford, 1963).
45. J. Eastman, Phys. Thin Films **10**, 167 (1978).
46. Lord Rayleigh, Phil. Mag. **14**, 350 (1907).
47. D. K. G. de Boer, Phys. Rev. B **49**, 5817 (1994).
48. D. K. G. de Boer, Phys. Rev. B **51**, 5297 (1995).
49. U. Klemradt (1994), *Charakterisierung epitaktischer GaAs/AlGaAs-Schichtsysteme mittels Streuung harter Röntgenstrahlung unter streifendem Einfall sowie Röntgenbeugung*, Ph.D. thesis, Institut für Festkörperforschung, KFA Jülich (1994), also available as: *Berichte des Forschungszentrum Jülich*, Vol. **2894**, (1994).
50. J. J. Thomson, Proc. Cam. Phil. Soc. **15**, Pt. 5 (1910).
51. M. Kuriyama, J. Phys. Soc. Japan **23**, 1369 (1967).
52. L. Tapfer and K. Ploog, Phys. Rev. B **33**, 5565 (1986).
53. V. S. Speriosu and T. J. Vreeland, J. Appl. Phys. **58**, 1591 (1984).
54. A. T. Macrander, Annu. Rev. Mater. Sci. **18**, 283 (1988).
55. J. Kervarec, J. Baudet, M. Caulet, P. Auvray, J. Y. Emery, and A. Regreny, J. Appl. Crystallogr. **17**, 196 (1984).
56. H. Kiessig, Ann. Phys. **10**, 715 (1931).
57. I. A. Blech and E. Meieran, J. Appl. Phys. **38**, 2913 (1967).
58. Q. Shen and S. Kycia, Phys. Rev. B **55**, 15791 (1997).
59. A. J. Steinfort, P. M. L. O. Scholte, A. Ettema, F. Tuinstra, M. Nielsen, E. Landemark, D.-M. Smilgies, R. Feidenhans'l, G. Falkenberg, L. Seehofer, et al., Phys. Rev. Lett. **77**, 2009 (1996).
60. V. M. Kaganer and K. H. Ploog, Phys. Rev. B **64**, 205301 (2001).
61. G. S. Pearson and D. A. Faux, J. Appl. Phys. **88**, 730 (2000).
62. J. R. Downes, D. A. Faux, and E. P. O'Reilly, J. Appl. Phys. **81**, 6700 (1997).
63. D. A. Faux and G. S. Pearson, Phys. Rev. B **62**, R4798 (2000).
64. I. Kegel, T. H. Metzger, A. Lorke, J. Peisl, J. Stangl, G. Bauer, K. Nordlund, W. V. Schoenfeld, and P. M. Petroff, Phys. Rev. B **63**, 035318 (2001).
65. I. Kegel, T. H. Metzger, A. Lorke, J. Peisl, J. Stangl, G. Bauer, J. M. Garcia, and P. M. Petroff, Phys. Rev. Lett. **85**, 1694 (2000).
66. I. Kegel, T. H. Metzger, P. Fratzl, J. Peisl, A. Lorke, J. M. Garcia, and P. M. Petroff, Europhys. Lett. **45**, 222 (1999).
67. S. Christiansen, M. Albrecht, H. P. Strunk, and H. J. Maier, Appl. Phys. Lett. **64**, 3617 (1994).

68. T. Wiebach, M. Schmidbauer, M. Hanke, H. Raidt, R. Köhler, and H. Wawra, Phys. Rev. B **61**, 5571 (2000).
69. T. Benabbas, P. Francios, Y. Androussi, and A. Lefebvre, J. Appl. Phys. **80**, 2763 (1996).
70. M. Grundmann, O. Stier, and D. Bimberg, Phys. Rev. B **52**, 11969 (1995).
71. C. Pryor, J. Kim, L. W. Wang, A. J. Williamson, and A. Zunger, J. Appl. Phys. **83**, 2548 (1998).
72. F. Buda, J. Kohanoff, and M. Parinello, Phys. Rev. Lett. **69**, 1272 (1992).
73. A. Saitta, F. Buda, G. Fiumara, and P. Giaquinta, Phys. Rev. B **53**, 1446 (1996).
74. M. Hanke (2002), *Streuung von Röntgenstrahlung an selbstorganisierten Halbleiter-Inselstrukturen*, Ph.D. thesis, Humboldt-Universität zu Berlin (Mensch und Buch Verlag, Berlin, 2002).
75. A. Guinier, *X-Ray Diffraction* (Freeman, San Francisco, 1968).
76. J. Stangl, T. Roch, V. Holý, M. Pinczolits, G. Springholz, G. Bauer, I. Kegel, T. H. Metzger, J. Zhu, K. Brunner, et al., J. Vac. Sci. Technol. B **18**, 2187 (2000).
77. I. Kegel, T. H. Metzger, J. Peisl, P. Schittenhelm, and G. Abstreiter, Appl. Phys. Lett. **74**, 2978 (1999).
78. V. Chamard, T. H. Metzger, E. Bellet-Amalric, B. Daudin, C. Adelmann, H. Mariette, and G. Mula, Appl. Phys. Lett. **79**, 1971 (2001).
79. K. Zhang, C. Heyn, W. Hansen, T. Schmidt, and J. Falta, Appl. Phys. Lett. **76**, 2229 (2000).
80. J. Tersoff, C. Teichert, and M. G. Lagally, Phys. Rev. Lett. **76**, 1675 (1996).
81. Q. Xie, A. Madhukar, P. Chen, and N. Kobayashi, Phys. Rev. Lett. **75**, 2542 (1995).
82. V. Holý, G. Springholz, M. Pinczolits, and G. Bauer, Phys. Rev. Lett. **83**, 356 (1999).
83. C. J. Huang, D. Z. Li, B. W. Cheng, J. Z. Yu, and Q. M. Wang, Appl. Phys. Lett. **77**, 2852 (2000).
84. P. Sutter, E. Mateeva-Sutter, and L. Vescan, Appl. Phys. Lett. **78**, 1736 (2001).
85. K. Brunner, J. Zhu, G. Abstreiter, O. Kienzle, and F. Ernst, Thin Solid Films **369**, 39 (2000).
86. G. Springholz, V. Holý, M. Pinczolits, and G. Bauer, Science **282**, 734 (1998).
87. G. Springholz, M. Pinczolits, P. Mayer, V. Holý, G. Bauer, H. H. Kang, and L. Salamanca-Riba, Phys. Rev. Lett. **84**, 4669 (2000).
88. I. Kegel, T. H. Metzger, J. Peisl, J. Stangl, G. Bauer, and D. Smilgies, Phys. Rev. B **60**, 2516 (1999).
89. M. Schmidbauer, T. Wiebach, H. Raidt, M. Hanke, R. Köhler, and H. Wawra, Phys. Rev. B **58**, 10523 (1998).
90. M. Renninger, Z. Kristallogr. **97**, 107 (1937).
91. J. M. Bijvoet and C. H. MacGillavry, Chemisch Weekblad **36**, 330 (1939).
92. W. N. Lipscomb, Acta Crystallogr. A **2**, 193 (1949).
93. R. Köhler, *Habilitationsschrift* (Humboldt Universität zu Berlin, 1993).
94. Z. G. Pinsker, *Dynamical Scattering of X-Rays in Crystals* (Springer, Heidelberg, 1978).
95. S. Tagaki, J. Phys. Soc. Japan **26**, 1239 (1969).

96. D. Taupin, Bull. Soc. Franc. Minér. Cryst. **87**, 469 (1964).
97. P. Mikulík and T. Baumbach, Phys. Rev. B **59**, 7632 (1999).
98. M. Tolan, W. Press, F. Brinkop, and J. P. Kotthaus, Phys. Rev. B **51**, 2239 (1995).
99. M. Tolan, G. König, L. Brüggemann, W. Press, F. Brinkop, and J. P. Kotthaus, Europhys. Lett. **20**, 223 (1992).
100. A. Authier, *Dynamical Theory of X-Ray Diffraction* (Oxford University Press, Oxford, 2001).
101. H. von Helmholtz, Crelles J. **100**, 213 (1886).
102. S. Dietrich and H. Wagner, Phys. Rev. Lett. **51**, 1469 (1983).
103. S. Dietrich and H. Wagner, Z. Phys. **56**, 207 (1984).
104. Y. Yoneda, Phys. Rev. **131**, 2010 (1963).
105. Y. Zhuang, U. Pietsch, J. Stangl, V. Holý, N. Darowski, J. Grenzer, S. Zerlauth, F. Schäffler, and G. Bauer, Physica B **283**, 130 (2000).
106. J. Stangl, V. Holý, T. Roch, A. Daniel, G. Bauer, J. Zhu, K. Brunner, and G. Abstreiter, Phys. Rev. B **62**, 7229 (2000).
107. V. Holý, T. Roch, J. Stangl, A. Daniel, G. Bauer, T. H. Metzger, Y. H. Zhu, K. Brunner, and G. Abstreiter, Phys. Rev. B **63**, 205318 (2001).
108. V. M. Kaganer, S. A. Stepanov, and R. Köhler, Phys. Rev. B **52**, 16369 (1995).
109. M. Rauscher, R. Paniago, H. Metzger, Z. Kovats, J. Domke, J. Peisl, H.-D. Pfannes, J. Schulze, and I. Eisele, J. Appl. Phys. **86**, 6763 (1999).
110. D. Grigoriev, M. Hanke, M. Schmidbauer, P. Schäfer, O. Konovalov, and R. Köhler, J. Phys. D **36**, A225 (2003).
111. E. A. Kondrashkina, S. A. Stepanov, R. Opitz, M. Schmidbauer, R. Köhler, R. Hey, M. Wassermeier, and D. V. Novikov, Phys. Rev. B **56**, 10469 (1997).
112. B. B. Mandelbrot, *The Fractal Geometry of Nature* (Freeman, New York, 1982).
113. S. F. Edwards and D. R. Wilkinson, Proc. R. Soc. London A **381**, 17 (1982).
114. M. Kardar, G. Parisi, and Y. C. Zhang, Phys. Rev. Lett. **56**, 889 (1986).
115. A.-L. Barabasi and H. E. Stanley, *Fractal Concepts in Surface Growth* (Cambridge University Press, Cambridge, 1995).
116. J. Krug and H. Spohn, *in: Solids far from Equilibrium* (Cambridge University Press, Cambridge, 1992).
117. D. G. Stearns, J. Appl. Phys. **71**, 4286 (1992).
118. E. Spiller, D. Stearns, and M. Krumrey, J. Appl. Phys. **74**, 107 (1993).
119. S. Pflanz and W. Moritz, Acta Crystallogr. A **48**, 716 (1992).
120. Z. H. Ming, A. Krol, Y. L. Soo, Y. H. Kao, J. S. Park, and K. L. Wang, Phys. Rev. B **47**, 16373 (1993).
121. R. Paniago, H. Homma, P. C. Chow, S. C. Moss, Z. Barnea, S. S. P. Parkin, and D. Cookson, Phys. Rev. B Rapid Commun. **52**, 17052 (1995).
122. Y. H. Phang, R. Kariotis, D. E. Savage, and M. G. Lagally, J. Appl. Phys. **72**, 4627 (1992).
123. J. Stangl, V. Holý, J. Grim, G. Bauer, J. Zhu, K. Brunner, G. Abstreiter, O. Kienzle, and F. Ernst, Thin Solid Films **357**, 71 (1999).
124. C. Giannini, T. Baumbach, D. Lübbert, R. Felici, L. Tapfer, T. Marschner, W. Stolz, N. Y. Jin-Phillipp, and F. Phillipp, Phys. Rev. B **61**, 2173 (2000).
125. S. Pflanz, H. L. Meyerheim, W. Moritz, I. K. Robinson, H. Hoernis, and E. H. Conrad, Phys. Rev. B **52**, 2914 (1995).

126. S. K. Sinha, M. K. Sanyal, S. K. Satija, C. F. Majkrzak, D. A. Neumann, H. Homma, S. Szpala, A. Gibeaud, and H. Morkoc, Physica B **198**, 72 (1994).
127. J. Stettner, L. Schwalowsky, O. H. Seeck, M. Tolan, W. Press, C. Schwarz, and H. von Känel, Phys. Rev. B **53**, 1398 (1996).
128. W. Press, M. Tolan, J. Stettner, O. H. Seeck, J. P. Schlomka, V. Nitz, L. Schwalowsky, P. Müller-Buschbaum, and D. Bahr, Physica B **221**, 1 (1996).
129. P. R. Pukite, C. S. Lent, and P. I. Cohen, Surf. Sci. **161**, 39 (1985).
130. M. Yoon, S. G. J. Mochrie, D. M. Zehner, G. M. Watson, and D. Gibbs, Phys. Rev. B **49**, 16702 (1994).
131. R. L. Headrick and J.-M. Baribeau, Phys. Rev. B **48**, 9174 (1993).
132. V. Holý, C. Giannini, L. Tapfer, T. Marschner, and W. Stolz, Phys. Rev. B **55**, 9960 (1997).
133. M. G. Friedel, C. R. Acad. Sci. Paris **157**, 1533 (1913).

3 Experimental Optimization

In this chapter, some basic experimental considerations are discussed. After the requirements, and the actual properties of synchrotron radiation sources are described, the most relevant experimental techniques and scattering geometries are introduced and compared with each other. A very crucial aspect is fast data acquisition. Since the diffuse scattering has to be recorded in all three dimensions of reciprocal space, this is only possible if multi-detection techniques are applied. This is implemented through the use of area detectors with a high dynamic range, i.e. high sensitivity and low background.

3.1 Requirements

The investigation of diffuse scattering from small structures in the nanometer and submicrometer regime places specific demands on the experimental setup. In this book we restrict ourselves to the *angle-dispersive methods*, where reciprocal space is probed by varying the directions of the incident and scattered wave vectors.[1]

The most relevant experimental demands can be summarized as follows:

- The essential part of the diffuse scattering of mesoscopic structures is found close to reciprocal-lattice points. This requires high angular resolution and a monochromatic X-ray beam.
- The X-ray wavelength should not be chosen too small, owing to the reduced q-space resolution and small atomic form factors. On the other hand, the X-ray wavelength cannot be chosen too large, since the Bragg condition requires $\lambda < 2d$, where d is the net-plane spacing. Moreover, air scattering – which is superimposed on the diffuse scattering from the sample – is particularly strong at low X-ray energies. Although air scattering is reduced by utilizing evacuated flight tubes between the sample and detector, the experiments are usually not performed entirely under vacuum conditions. Therefore, the optimum wavelength λ is in the range of 1–2 Å.

[1] Energy-dispersive X-ray diffraction (e.g. [1, 2]) and energy-dispersive reflectivity [3] are also frequently applied. Here, a white beam is used, and the scattered intensity at fixed angles is detected as a function of X-ray energy.

- Since the penetration depth of X-rays in a semiconductor material is of the order of several tens of micrometers, the diffusely scattered signal from mesoscopic structures with sizes in the nanometer or submicrometer regime is usually rather weak. Moreover, the diffuse scattering is widely distributed in reciprocal space. For that reason, a very intense X-ray source is necessary.
- The shape and strain field of mesoscopic structures leads to a (complicated) three-dimensional (3D) intensity distribution of diffuse scattering. Therefore, one-dimensional rocking curves that integrate over a certain range in reciprocal space do not contain sufficient information. Instead, the diffuse scattering has to be recorded in 3D or at least in several 2D-sections. As a consequence of the required 3D data acquisition, sufficient resolution is needed in all three directions of reciprocal space. In addition, sufficient counting statistics require fast data acquisition and highly sensitive detectors. Therefore, sophisticated multidetection techniques are necessary. These will be described in Sect. 3.5.

3.2 Synchrotron Radiation

The rather weak signal of X-ray diffuse scattering from mesoscopic structures requires a very bright and collimated X-ray beam. Both of these requirements, namely high collimation along with a high photon flux, are intrinsically satisfied by highly brilliant third-generation synchrotron radiation. Besides brilliance – which is defined as the number of photons per second emitted into a certain solid angle normalized to the source area and a fixed energy bandwidth – synchrotron radiation covers a large spectral range. Moreover, it is horizontally linearly polarized for light emitted within the plane of the electron orbit, whereas it is elliptically polarized above and below the orbit. Last not least, it exhibits a periodic time structure with bunch lengths in the subnanosecond range. Nowadays, synchrotron radiation is widely used in many fields of physics, chemistry, and biology. For extended review articles, see [4, 5, 6].

3.2.1 Beam Collimation

Here, we shall focus on the collimation conditions of synchrotron radiation sources. At a bending magnet, the vertical divergence is inherently given by

$$\Delta\theta_{\mathrm{v}} = \gamma^{-1} , \tag{3.1}$$

where $\gamma = E/E_0$ is the ratio of the electron (or positron) energy E to the electron rest energy E_0. Typical values are $\gamma^{-1} = 0.85 \times 10^{-4}$ (ESRF) and $\gamma^{-1} = 1.14 \times 10^{-4}$ (HASYLAB). In the horizontal plane, synchrotron radiation from a bending magnet exhibits a fan-like angular distribution which can be reduced geometrically by inserting aperture slits in front of the sample.

When wiggler radiation is used, the horizontal divergence (full width at half maximum, FWHM) is reduced to [7, 8]

$$\Delta\theta_{\mathrm{h}} = K\gamma^{-1} , \tag{3.2}$$

while the vertical divergence is still given by (3.1). The deflection parameter K is a measure of the maximum horizontal deflection of an electron when passing through a wiggler or undulator. It is defined as

$$K = \frac{eB_0\lambda_{\mathrm{u}}}{2\pi mc_0} = 0.934\lambda_{\mathrm{u}}(\mathrm{cm})B_0(\mathrm{T}) , \tag{3.3}$$

where B_0 is the on-zone magnetic field and λ_{u} the period length [7, 8]. For typical values for wigglers ($K \approx 10$), the horizontal collimation is of order $\Delta\theta_{\mathrm{h}} \approx 1\,\mathrm{mrad}$.

In order to optimize the flux at the sample, many experimental stations use optical elements (monochromators or mirrors) that horizontally focus the X-ray beam onto the sample position. The horizontal divergence is then in the order of several milliradians. If one is aiming to perform 3D reciprocal-space mapping, this value is often too high. A reduction of the divergence can be achieved by placing aperture slits in front of the focusing element. This is, however, only possible at the expense of photon flux.

When undulator radiation ($K \approx 1$) is used, the divergences are further reduced [7, 8], and both the vertical and the horizontal divergences are given by

$$\Delta\theta_{\mathrm{h}} = \Delta\theta_{\mathrm{v}} = \sqrt{\frac{1 + K^2/2}{2nN}}\gamma^{-1} , \tag{3.4}$$

where N is the number of magnet periods and n is the index of the harmonic. This is a substantial reduction compared with the natural collimation γ^{-1}. Typical values are of order $\Delta\theta_{\mathrm{v}} = \Delta\theta_{\mathrm{h}} \approx 10\,\mu\mathrm{rad}$. For these small values the finite divergences of the electron beam, which are of the same order of magnitude but are different for the horizontal and vertical directions, have to be taken into account.

3.2.2 Energy Resolution

The excellent vertical collimation of synchrotron radiation can be exploited to extract a small wavelength band out of the large spectral range. Usually, a double-crystal monochromator is employed, leading to a typical resolution of $\Delta E/E = \Delta\lambda/\lambda \approx 10^{-5}$–$10^{-4}$, depending on the source collimation and the type of monochromator.

3.3 Coherence of X-Rays

An X-ray diffraction experiment always requires sufficient coherence of the X-ray source. In the previous theoretical chapter the coherence has been taken for granted and the incoming X-ray beam has, implicitly, been treated as a perfectly monochromatic infinite plane wave of the form $E(\boldsymbol{r},t) = E_0 e^{i(\boldsymbol{k}\cdot\boldsymbol{r}-\omega t)}$. This is equivalent to an infinite coherence length of the X-ray beam and is obviously an idealization. A "real" X-ray beam always deviates from this idealized picture since it neither is perfectly monochromatic nor has a well-defined direction of propagation. Both properties can be quantitatively expressed through the longitudinal (or temporal) coherence length $L_{\|}$ and the transverse (or spatial) coherence length $L_{\perp}$.

Coherence lengths can be easily measured by Fraunhofer diffraction from a pinhole. However, the introduction of other optical elements such as collimators, crystal analyzers, and focusing mirrors into the X-ray beam may alter the coherence properties of the X-ray beam. In that case the simple relations between the coherence lengths and the source properties given below may not be valid. Effects due to the finite vertical divergence of the beam are also neglected here, and the coherence is considered exclusively in the scattering plane. For a detailed description in the case of inserted optical elements, we refer the reader to the excellent work of Kaganer et al. [9].

3.3.1 Coherence Lengths

The most critical parameter in a coherent diffraction experiment is the spatial (or transverse) coherence length $L_{\perp}$. A perfectly monochromatic plane wave has a constant phase everywhere in a plane perpendicular to its wave vector $\boldsymbol{k}$. Two monochromatic waves propagating in directions differing by an angle $\Delta\theta$ will have a phase difference of 2π within this plane at a distance $L_{\perp}$, which is called the coherence length and is given by

$$L_{\perp} = \frac{2\pi}{k\,\Delta\theta} = \frac{\lambda}{\Delta\theta}\,. \tag{3.5}$$

Here $\Delta\theta$ is the angle that the X-ray source is seen to subtend when one looks back from a fixed position of the sample to the source, i.e. it is given by the horizontal and vertical apertures of the X-ray source. For a Gaussian distribution, (3.5) can be expressed as[2]

$$L_{\perp} = \frac{\lambda R}{s}\left(\frac{2\ln 2}{\pi}\right)^{1/2}, \tag{3.6}$$

[2] This relationship depends on the geometry of the X-ray source. Equation (3.6) holds for a Gaussian distribution. For a rectangular distribution with a base width a, the relation is $L_{\perp} = \lambda R/a$, whereas $L_{\perp} = (\lambda R/\varrho)(2/\pi)$ holds for a circular distribution with radius ϱ.

where R is the distance between the X-ray source and the sample and s is the FWHM of the X-ray source. If we use $\lambda = 0.154\,\mathrm{nm}$, $R = 40\,\mathrm{m}$, and $s = 1\,\mathrm{mm}$ (typical values for HASYLAB), we obtain $L_\perp \approx 4\,\mu\mathrm{m}$. Owing to the smaller source size, substantially larger coherence lengths can be achieved at ESRF: $R = 46\,\mathrm{m}$ and $s = 0.02\,\mathrm{mm}$ give $L_\perp \approx 200\,\mu\mathrm{m}$.

The spatial coherence depends solely on geometric factors. In contrast, the longitudinal or temporal coherence is a consequence of the spectral purity and is given by

$$L_{\parallel} = \frac{\lambda^2}{2\,\Delta\lambda}\,, \tag{3.7}$$

where $\Delta\lambda$ is the bandwidth of the wavelength λ. Typical values are in the range of a few micrometers.

3.3.2 Grazing-Incidence Conditions

The total coherence lengths at the sample are given by the projections of $L_\perp$ and $L_\parallel$ onto the sample surface (Fig. 3.1), and are given by

$$L_\perp^{\mathrm{eff}} = \frac{L_\perp}{\sin\alpha_{\mathrm{i}}}\,, \tag{3.8a}$$

$$L_\parallel^{\mathrm{eff}} = \frac{L_\parallel}{\cos\alpha_{\mathrm{i}}}\,. \tag{3.8b}$$

In a grazing-incidence scattering geometry the angle of incidence is in the range of a fraction of a degree, resulting in a very large transverse coherence length. Several groups [10, 11] have proved experimentally that this idea is correct, and they were able to detect an effective transverse coherence length of up to $L_\perp^{\mathrm{eff}} = 60\,\mu\mathrm{m}$.

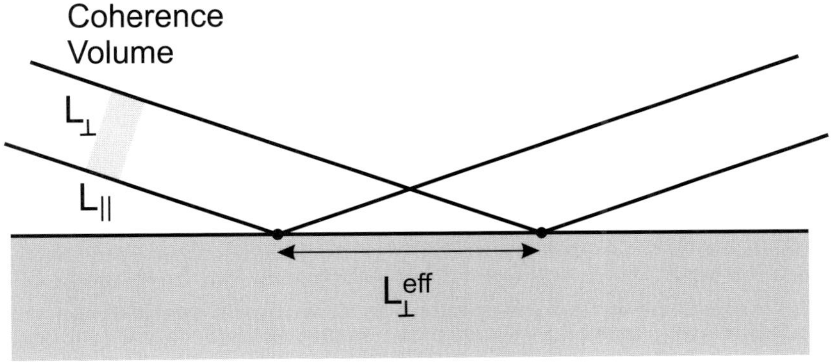

Fig. 3.1. The projection of the coherence volume, defined by the transverse and longitudinal coherence lengths $L_\perp$ and $L_\parallel$, respectively, onto the sample surface defines an effective coherence length $L_\perp^{\mathrm{eff}}$

3.3.3 Influence of a Position-Sensitive Detector on the Measured Coherence Lengths

The coherence lengths introduced in Sect. 3.3.1 relate to the coherently illuminated length across the sample surface. When, however, a linear position-sensitive detector (PSD) or an area detector is used, where different channels of the detector correspond to different scattering angles (see Sect. 3.5), the detector resolution has to be considered. The transverse coherence length of the incoming beam was determined by calculating the effective aperture of the source when looking from a fixed point of the sample back to the source. Analogously the detector aperture, that is, the effective aperture seen when looking from a fixed point of the sample at the smallest resolution element of the detector (which plays the role of an effective source size) has to be taken into account. A typical distances between detector and sample is $R = 800\,\mathrm{mm}$, and the minimum size of a resolution element is $s = 80\,\mu\mathrm{m}$ (or more). From these values and (3.6), the experimental limitation of the actual transverse coherence when a PSD is used can be estimated as $L_\perp \approx 1\,\mu\mathrm{m}$, which is considerably smaller than the source-related coherence lengths of synchrotron radiation. In other words, although the sample is illuminated coherently, the effective coherence length is smaller and is limited by the angular resolution of the position-sensitive detector.

3.3.4 Influence of Limited Coherence on the X-Ray Diffraction Pattern

The interference function $G(\boldsymbol{q})$ defined in (2.52) is exact for an X-ray source with perfect coherence. It is given as a coherent superposition of the scattered partial waves from the mesoscopic structures of the illuminated sample. Consequently, if a diffraction experiment is performed with highly coherent X-rays, i.e. the illuminated spot size at the sample is not much larger than the coherence length, the observed intensity will depend on the relative phases of the scattering from all scatterers of the illuminated sample. In the case of an absolutely perfect sample, sharp Bragg peaks evolve. However, any disorder in the sample structure – for example surface roughness, domain structure in a polycrystal, or disorder created during phase transitions – will lead to a characteristic speckle pattern in the experimental intensity distribution [12, 13, 14, 15, 16, 17].[3]

If, however, the illuminated spot size at the sample is much larger than the coherence length, the calculation of the interference function $G(\boldsymbol{q})$ in (2.52) has to be modified so that the summation is performed over the coherently illuminated volume of the sample only. $G(\boldsymbol{q})$ has then to be statistically

[3] Historically, speckle and related techniques such as dynamical light scattering [18, 19] have been used within the visible-light region to study diffusion and critical phenomena in fluids.

averaged over all possible island configurations, i.e. the intensities of different coherently scattering parts of the sample have to be summed incoherently. This ensemble averaging is expressed by the brackets $\langle \ldots \rangle$ in (2.53).

When the coherence is reduced, the number of addends in (2.53) is reduced, leading to a significant broadening of the peaks. However, strong disorder in the sample itself leads to peak broadening even in the case of perfect X-ray coherence. Therefore, the two effects of limited coherence of the X-ray beam and limited coherence of the sample are in some sense equivalent, and coherence length and correlation length are closely related to each other. The measured peak widths are always a mixture of the results of these two effects, and the effective peak width is determined by the smaller of the coherence or the correlation length. Therefore, in order to measure the true properties of the sample, it is necessary that the coherence of the incident X-ray beam is sufficiently large. A good theoretical description of the effects of partial coherence on the scattering of X-rays by matter is given by Sinha et al. [20].

3.4 Scattering Geometries: the Extended Ewald Sphere

It has been shown in Chap. 2 that the scattered intensity can be expressed in terms of the scattering vector $\boldsymbol{q} = \boldsymbol{k}^{\mathrm{f}} - \boldsymbol{k}^{\mathrm{i}}$, where $\boldsymbol{k}^{\mathrm{i,f}}$ are the wave vectors of the incoming and scattered waves, respectively. Following the definition of the angles $\alpha_{\mathrm{i,f}}$ and $\theta_{\mathrm{i,f}}$ in Fig. 3.2, the components of the scattering vector can be expressed as

$$
\begin{aligned}
q_x &= k(\cos\alpha_{\mathrm{f}}\cos\theta_{\mathrm{f}} - \cos\alpha_{\mathrm{i}}\cos\theta_{\mathrm{i}}) \; , \\
q_y &= k(\cos\alpha_{\mathrm{f}}\sin\theta_{\mathrm{f}} + \cos\alpha_{\mathrm{i}}\sin\theta_{\mathrm{i}}) \; , \\
q_z &= k(\sin\alpha_{\mathrm{f}} + \sin\alpha_{\mathrm{i}}) \; . \qquad (3.9)
\end{aligned}
$$

Here $\alpha_{\mathrm{i,f}}$ are the glancing angles of incidence and exit with respect to the surface (x–y plane), and $\theta_{\mathrm{i,f}}$ are the in-plane angles. Often, the experimentally

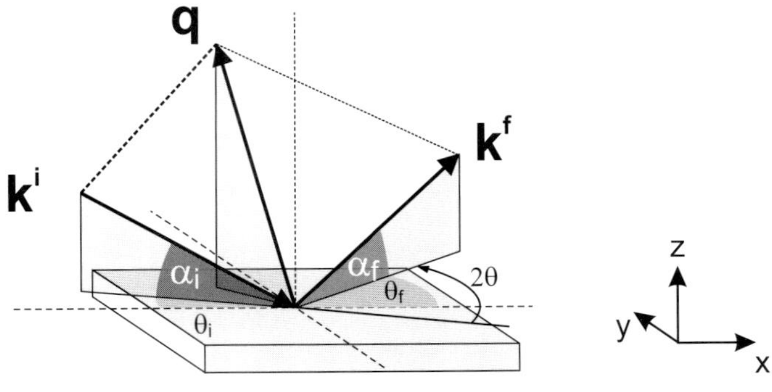

Fig. 3.2. Definition of the scattering vector $\boldsymbol{q}$ by the glancing angles α_{i}, α_{f} and in-plane angles θ_{i}, θ_{f}, 2θ

more convenient *in-plane scattering angle* $2\theta = \theta_f + \theta_i$ is used instead of θ_f (see Fig. 3.2).

Reciprocal space can be probed by choosing appropriate directions of the incident and scattered beam, as shown in Fig. 3.3. There are, however, some restrictions on the regions in reciprocal space that are accessible in a diffraction experiment. As only elastic interactions are considered here, the length of the wave vector is conserved, i.e. $|\boldsymbol{k}^i| = |\boldsymbol{k}^f| = k = 2\pi/\lambda$. Since the maximum length of the scattering vector is given by $|\boldsymbol{q}| = 2k$, the accessible range of allowed scattering vectors is restricted to the area inside a hemisphere of radius $2k$, which is called the *extended Ewald sphere*. The samples discussed here are thin mesoscopic structures that are grown on a rather thick substrate. When semiconductor substrate materials are investigated at wavelengths in the interval of 1–2 Å, the X-ray beam cannot penetrate through the sample. Therefore, reflection geometries (in contrast to transmission geometries) have to be applied. This restricts the allowed area of reciprocal space to areas that are outside the Laue zones, as illustrated in Fig. 3.3.

The Laue zones are aligned along the x axis, which is defined as the intersection of the scattering plane and the surface. Since any movement in reciprocal space may change the scattering plane, the extended Ewald

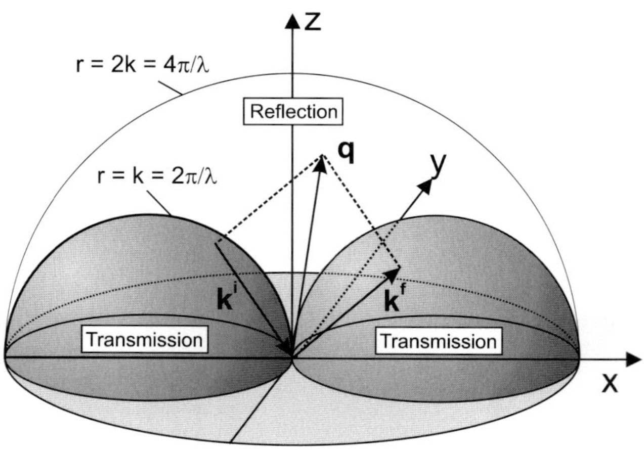

Fig. 3.3. Schematic view of the region of reciprocal space that is accessible by elastic scattering. The basal plane is parallel to the sample surface, with the normal vector pointing outwards in the z direction. The scattering vector $\boldsymbol{q} = \boldsymbol{k}^f - \boldsymbol{k}^i$ must be located inside the large hemisphere (radius $r = 2k = 4\pi/\lambda$), since its length is limited by $q \leq 2k$. The *gray-shaded* Laue zones (radius $r = k = 2\pi/\lambda$) represent transmission geometries. Since neither the incoming nor the diffracted beam can penetrate through the sample, the Laue zones are inaccessible. By systematically varying the directions of the incident and scattered wave vectors $\boldsymbol{k}^i$ and $\boldsymbol{k}^f$, respectively, one can use the scattering vector $\boldsymbol{q}$ to map out different volumes of reciprocal space

construction has to be – strictly speaking – redefined for each scattering vector $\boldsymbol{q}$. However, as long the movement of $\boldsymbol{q}$ in reciprocal space during a scan is only small, this quasi-stationary model is still a good approximation.

3.5 Multidetection Techniques

3.5.1 Basic Principle

Precise measurements of both the incident angle and the scattering angle require a *triple-crystal*[4] setup, the principal advantage of which is the ability to map reciprocal space with extremely high resolution. As pointed out above, a well-defined direction and length of the incident wave vector are guaranteed if one uses a collimated, monochromatic X-ray beam. The direction of the scattered beam is then analyzed by a crystal analyzer or, when only medium resolution is required, by a collimating slit system. The actual choice depends on the scattering geometry and, therefore, on the required resolution, dynamical range, and area of interest in reciprocal space. However, the corresponding large data collection time represents a serious disadvantage and makes this technique not feasible for recording the three-dimensional intensity distribution of diffuse scattering, even if a very intense X-ray source is available. The scanning of, for example, $100 \times 100 \times 100 = 10^6$ data points where each point is recorded in a time of only 1 s requires a total data collection time of several hundreds of hours.

The difficulty of long data collection times that arises when a triple-axis diffractometer is used can be overcome by utilizing modern area detectors or, at least, a linear position-sensitive detector. This idea was first suggested by Mathieson [21], and its application to epitaxial materials was discussed by Lee et al. [22]. At a sufficiently small spot size at the sample and sufficiently good spatial resolution, an area detector offers *parallel determination* of the scattering angles α_f and 2θ (see Fig. 3.4). This multidetection technique substantially reduces the data acquisition time, and a 2D mapping (using a linear PSD) or a 3D mapping (using an area detector) of reciprocal space can then be performed in a very short time, similar to that for a one-dimensional rocking scan.

For a spot size in the range of 200 μm and a distance between the sample and the detector of about 1000 mm, the angular resolution of the diffracted beam is of order $\Delta\theta = 2 \times 10^{-4}$ rad. This intermediate resolution – though definitely worse than the high resolution provided by a crystal analyzer – often turns out to be sufficient for the analysis of diffuse scattering.

[4] The term *triple-crystal* refers to the number of optical components that are used in the experiment: (1) a monochromator–collimator crystal, (2) the sample, and (3) an analyzer crystal. When the analyzer stage is left out, the term *double-crystal* is used.

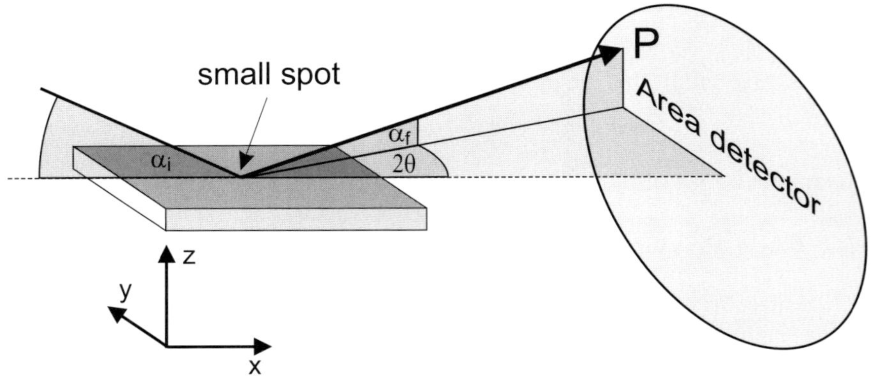

Fig. 3.4. Principle of multidetection in an X-ray scattering experiment. For a small spot size at the sample and a sufficiently good spatial resolution of the detector, each point P on an area detector corresponds to unique values of $\alpha_{\rm f}$ and 2θ. The diffuse scattering at different values of these angles is probed in a *parallel* measurement without scanning the sample (fixed $\alpha_{\rm i}$ and $\theta_{\rm i}$). Alternatively, a linear position-sensitive detector can be used, aligned along either the z or the y direction. The angular resolution depends on the spot size at the sample, the spatial resolution of the detector, and the distance between the detector and the sample

The need to probe the diffuse scattering over a large dynamic range close to strong coherent beams (e.g. Bragg reflections or the specular beam) requires a "clean" incident beam profile. On the other hand a small spot size of typically 50–200 µm at the sample is necessary. Therefore, high-quality slits have to be used, and a setup consisting of sequential slits, the first acting as a beam size limitation and the second suppressing the scattering from the first slit, is most advantageous. It is very advantageous to use selective attenuators (e.g. thin metal wires) in front of the area detector to strongly suppress the coherent beams and thus avoid detector overflow or damage.

3.5.2 Area Detectors

The use of an area detector for 3D reciprocal-space mapping requires the satisfaction of a variety of experimental conditions. The most important are:

- A high dynamic range of typically 10^4–10^5 is needed. This implies high quantum efficiency and low background.
- The detector linearity has to be guaranteed over the entire dynamic range.
- The spatial resolution has to be better than 100 µm (FWHM) without crosstalk.
- Fast readout has to be guaranteed.
- The available sensitive area should be in the region of 50 × 50 mm^2. This would correspond to an aperture of about 3° at 1000 mm distance from the sample.

Below, the most popular area detectors are briefly introduced and discussed with respect to their applicability as sensors for 3D reciprocal-space mapping of diffuse scattering.

Gas-Filled Wire Detectors. These detectors – here called simply position-sensitive detectors – are based on the simple mechanism of ionization in a noble gas under high pressure.[5] They are thus similar to a simple Geiger–Müller detector; however, they are used in the proportional counting mode. After absorption of an X-ray photon, the resulting ions and electrons are accelerated to the cathode and anode, respectively. This is done by applying a high voltage between the housing (cathode) and a stretched thin metal wire (anode), which is placed in the center of the housing. The position of the impinging electron pulse on the wire is determined either by measuring the time delay between the pulses at the two ends of the wire or by measuring difference in the heights of the two pulses. In the first case a low-resistance wire (e.g. gold or platinum wire) is used, whereas the second case requires a high-resistance wire, e.g. carbon-coated quartz. In both cases the spatial resolution is of order 100 μm.

The main advantage of a gas-filled wire detector is its rather high quantum efficiency (typically 50%) and its high linearity and spatial resolution, which fits well to the experimental conditions. Since the detector is used in the proportional mode, there is some (poor) energy resolution that can be used to extract the true experimental signal from the high-energy background (higher harmonics) and the low-energy thermal noise. The disadvantages are the restricted dynamic range (not more than 10 000 cps) and considerable crosstalk between neighboring channels (on a logarithmic scale!).

Gas-filled multi-wire detectors based on the above principles, in a crossed 2D arrangement [23], are preferentially used in SAXS experiments, where the detector can be placed far from the sample (4–20 m). Therefore, the spatial resolution of the detector itself need not be excellent, and typical values are in the order of 1 mm (typically 128 wires covering a width of 10 cm) or larger. Owing to the limited distance between the sample and the detector, these multi-wire detectors are generally not suitable for the investigation of diffuse scattering from mesoscopic structures. Moreover, these detectors are very expensive.

Image Plates. Image plate detectors [24] are based on the principle of the activation of photoelectrons into metastable F-centers of Eu-doped barium halides after illumination by a single X-ray photon. The positions of these metastable states can be read out by laser scanning. This is carried out by illuminating the image plate with green laser light, which induces relaxation of the electrons into the ground state, accompanied by emission of blue light.

[5] The choice of the noble gas depends on the X-ray energy. At about 10 keV argon is used, whereas for higher energies xenon is favored. Usually, the noble gas contains a small fraction of quenching gas (e.g. 5% CH_4) that discharges the charge cloud after it is created.

With this technique, sufficient resolution, typically larger than 100 μm, is achieved and the image plate is linear and exhibits high quantum efficiency along with a high dynamic range ($\approx 10^5$). However, the restriction to *off-line* readout makes this detector system very unpractical in its handling on a diffractometer, since 3D mapping requires a sequence of many illuminations in a rather short time.

CCD Detectors. CCDs (charge-coupled devices) [25] are modern detector systems which have recently become very attractive as X-ray detectors. In most experimental applications, CCD area detectors are used for X-ray imaging (tomography and topography) or structure analysis (by the Laue or Debye–Scherrer technique). For imaging techniques, very good spatial resolution is necessary, but in most cases the dynamic range need not be so large. For structure analysis, a large sensitive area is necessary, while the dynamic range and resolution need not be very large. In contrast, when a CCD detector is used for probing diffuse scattering, the most critical requirement is a high dynamic range of at least five orders of magnitude. This means that the intensity scale within a single frame must span a range of 10^5. This requires a high signal-to-noise ratio and high sensitivity. On the other hand, the required angular resolution of typically 10^{-4} rad implies a typical spatial resolution of the detector of better than 100 μm.

Basically, there are two different types of CCD detectors. In the first type, *direct* systems, the X-ray photons impinge directly on the CCD chip. In this case the detector is rather small and easy to handle within the experiment. The setup consist of two elements only, the CCD sensor and a mechanical shutter. A single 8 keVphoton produces about 2500 electron–hole pairs in silicon, and the corresponding signal is easy to detect. However, the interaction of X-rays with silicon is rather weak ($1/\mu = 70$ μm at 8 keV), and only a small fraction of the X-rays is absorbed in the thin (a few micrometers) depletion layer of the chip. Consequently, a direct CCD system exhibits only a small quantum efficiency. Therefore, *indirect* systems have been developed where the X-ray photon is first converted into visible light. For that purpose, a polycrystalline phosphor screen is often used as a scintillator. As a consequence of using visible light that is easily absorbed in the depletion layer, a rather high quantum efficiency can be achieved, up to single-photon counting. Often a fiber-optic element (*taper*) is inserted between the scintillator and the CCD in order to allow for images larger than the actual size of the CCD chip. This, however, leads to significant losses and warpage of the image. Another serious disadvantage of an indirect system is the high weight and the bulky dimensions of the detector.

A very critical point is the noise of the CCD, since it limits the true dynamic range of the detector. In order to minimize the dark current, the CCD chip has to be cooled. This is usually performed by a multistage Peltier cooling element working at approximately $-70°\mathrm{C} \ldots -40°\mathrm{C}$. To a first approximation, the dark current behaves like a statistical quantity, and is of the order of $0.01\,\mathrm{e\,pixel^{-1}\,s^{-1}}$ in commercial systems.[6] Therefore, the signal-to-noise ratio can be efficiently enhanced by sufficiently large illumination times. Besides the dark current, there is a constant amount of noise that is generated by the electronics during readout of the CCD. The manufacturers of commercial systems claim that the readout noise is equivalent to 1 X-ray photon $\mathrm{pixel^{-1}}$. However, fast readout is required. Unfortunately, fast readout is complementary to low background, since the readout noise scales inversely with readout time. Here, a reasonable compromise between speed and dynamic range has to be found. The mean background can be partly eliminated by subtracting a dark-current frame that is created with the shutter closed. The achievable dynamic range is of order 15–17 bits, i.e. 33 000–130 000 photons $\mathrm{pixel^{-1}}$.

3.6 Scattering Techniques

3.6.1 Coplanar High-Resolution Diffraction

The term *high-resolution X-ray diffraction* (HRXRD) is used for all diffraction geometries where the reciprocal-lattice vector fulfills the condition $\boldsymbol{h} \neq 0$. Usually, the term *high-resolution* is applied when a triple-crystal setup is used. Nevertheless, the resolution achieved by area or line detectors is still high enough to justify describing the scattering as highly resolved. *Coplanar* scattering means that the scattering plane (defined by $\boldsymbol{k}^{\mathrm{i}}$ and $\boldsymbol{k}^{\mathrm{f}}$) coincides with the plane of incidence (defined by $\boldsymbol{k}^{\mathrm{i}}$ and the surface normal vector $\boldsymbol{z}$). As a consequence of this scattering geometry, only the x–z plane of the extended Ewald sphere in Fig. 3.3 has to be considered. The corresponding section is shown in Fig. 3.5. Also shown in this picture is the functionality of a linear position-sensitive detector. Using a PSD in the plane of Fig. 3.5 is equivalent to a scan in reciprocal space along the thick line at the tip of $\boldsymbol{q}$.

The term *coplanar* is also used in the sense of probing the 3D intensity distribution in the proximity of a coplanar reciprocal-lattice point, i.e. the coherent Bragg reflection is exactly coplanar, but the diffuse scattering may deviate slightly from the exact coplanar condition. This means that the in-plane scattering vector and the corresponding in-plane angles do not vanish, but are very small. These conditions apply also when a two-dimensional area detector is used instead of a linear PSD. Similarly to the region marked by the curved thick line in Fig. 3.5, an area detector would correspond to multitude of detector scans in an analogous curved plane.

Usually, one has to differentiate between *symmetrical* and *asymmetrical* reflections. In the first case, the diffuse scattering is investigated around

[6] The technology leaders are Mar Research (www.marresearch.com) and Bruker AXS (www.brukeraxs.com). Parts of the company Enraf-Nonius have been taken over by Bruker AXS and are now called Bruker Nonius (www.nonius.com).

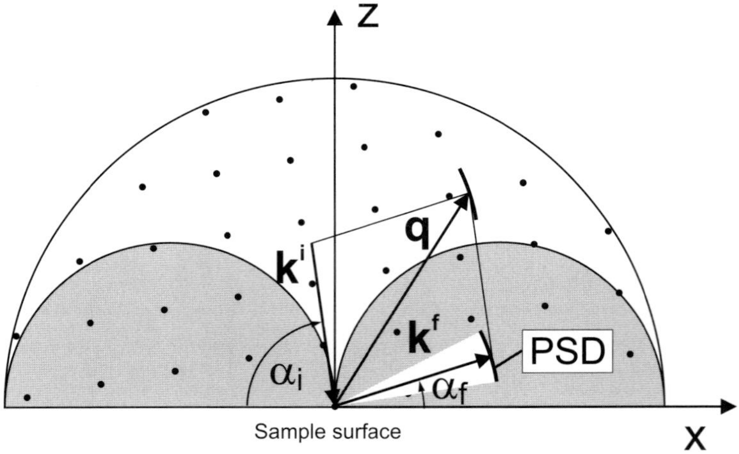

Fig. 3.5. Schematic illustration of coplanar HRXRD scattering geometry. A 2D section through the extended Ewald sphere (Fig. 3.3) at $q_y = 0$ is shown, which is chosen so as to contain the wave vectors $\boldsymbol{k}^{\mathrm{i}}$ and $\boldsymbol{k}^{\mathrm{f}}$ of the incident and the scattered wave. The sample normal vector $\boldsymbol{z}$ lies in the scattering plane (coplanar scattering). A linear PSD records simultaneously the intensity distribution along α_{f}. The α_{f} distribution corresponds to the region in reciprocal space that is marked by the *solid line* at the tip of $\boldsymbol{q}$. A two-dimensional detector would correspond to a multitude of scans in an analogous curved plane. Reciprocal-lattice points are marked by *black dots*

a reciprocal-lattice point with a vanishing x component ($h_x = 0$). This is equivalent to coherent diffraction from net planes that are oriented parallel to the sample surface. From a symmetrical reflection, information can be obtained about lattice strains in the vertical direction only, and lateral strains are not accessible. This circumstance is rather advantageous in that strain information can be obtained in the direction along $\boldsymbol{q}$ (the radial direction), while the shape, possible lateral correlation, and local lattice tilts can be observed perpendicular to $\boldsymbol{q}$ (the angular direction). This facilitates the interpretation of diffuse scattering. In the other case of an *asymmetrical* reflection, the intensity is investigated in the vicinity of a reciprocal-lattice point with a nonvanishing h_x component. This is equivalent to diffraction from net planes that are inclined with respect to the surface. For an asymmetrical reflection, a discrimination into strain, tilt, and shape effects is not possible anymore.

Generally, a complete understanding of the scattering pattern is only possible if it is accompanied by theoretical simulations. The diffuse-scattering simulations can be usually performed in the kinematical approximation as long as the incident and exit angles $\alpha_{\mathrm{i,f}}$ are large compared with the critical angle of total external reflection.

3.6.2 Grazing-Incidence Diffraction

This is similar to HRXRD, i.e. $\boldsymbol{h} \neq 0$, but now the scattering plane is not identical to the plane of incidence but is chosen nearly parallel to the crystal surface. In this case the incident and scattered X-rays hit the surface at very grazing angles (Fig. 3.6). As a consequence of this extremely noncoplanar geometry, good collimation of the incident beam is required in the directions parallel and perpendicular to the plane of incidence, i.e. both of the angular divergences $\Delta\alpha_{\mathrm{i}}$ and $\Delta\theta_{\mathrm{i}}$ have to be sufficiently small. Moreover, the grazing incidence restricts the beam height accepted by the sample to less than a few hundreds of micrometers. Therefore, triple-crystal GID has not been practicable in normal laboratories for the measurement of (weak) 3D diffuse scattering.[7]

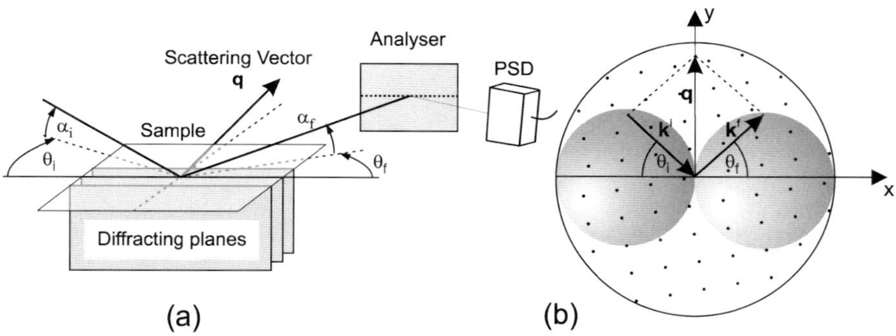

Fig. 3.6. (**a**) Schematic illustration of a triple-crystal setup for GID measurement. A collimated, monochromatic X-ray beam hits the sample at a grazing angle of incidence. It is then diffracted from net planes that are oriented approximately perpendicular to the surface so that the diffracted beam can leave the sample surface at a grazing exit angle. A crystal analyzer provides high resolution in the sample plane, and a PSD records simultaneously the intensity distribution as a function of α_{f}. In some cases the analyzer can be replaced by a collimating slit system. (**b**) Scattering geometry in the extended Ewald construction (in-plane 2D section)

The use of a 2D area detector is not advantageous for GID, since the diffracted beam is rather broad in the lateral direction, resulting in a poor angular resolution in this direction. Therefore, a linear position-sensitive detector oriented perpendicular to the surface is often used to record different

[7] The situation may have changed recently. In recent years, powerful focusing multilayer X-ray mirrors (parabolic Göbel mirrors [26]) have been developed. These mirrors, which are also available in a crossed-coupling construction, combined with high-power rotating-anode generators, yield typical intensities of the monochromatized primary beam of the order of several times 10^8 photons s^{-1}. A similar photon flux, though with a substantially smaller beam size, is achieved with bending magnets at HASYLAB.

values of the exit angle α_f with respect to the surface (Fig. 3.6), while a crystal analyzer selects the in-plane angle θ_f. With that geometry (triple-crystal GID), a 3D mapping of reciprocal space is possible. When only medium resolution is necessary, a collimation slit system is often used instead of an analyzer stage. Compared with HRXRD, refraction effects are important in the GID geometry. Since the index of refraction is smaller than one for X-rays, total external reflection occurs below a critical glancing angle α_c, and the X-ray penetration depth shrinks to a few nanometers (Fig. 2.2). Slightly above the critical angle, the penetration depth is about a factor of hundred larger. Therefore, GID enables one to tune the information depth from a few nanometers up to a few hundreds of nanometers. This in demonstrated in Fig. 3.7. Since mesoscopic structures are often buried at comparable depths, the use of GID leads to an enhanced sensitivity compared with HRXRD.

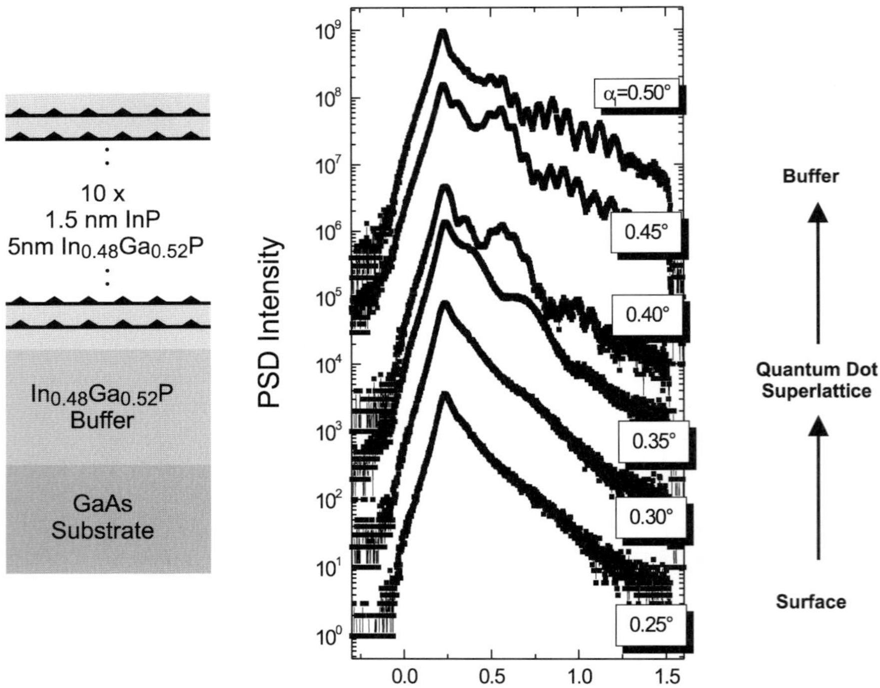

Fig. 3.7. Demonstration of depth resolution in grazing-incidence diffraction. The GID 220 reflection of an InP/$In_{0.48}Ga_{0.52}P$ quantum dot superlattice grown on GaAs(001) is shown. The intensity distribution along α_f was recorded by using a PSD. By tuning α_i, different depths of the sample can be visualized. At $\alpha_i < \alpha_c = 0.28°$, the superlattice structure is not visible. At $\alpha_i \geq \alpha_c$, the quantum dot superlattice satellite peaks appear, whereas at $\alpha_i \geq 0.40°$ thickness oscillations from the thick $In_{0.48}Ga_{0.52}P$ buffer layer are additionally superimposed

Since the vertical component of $\boldsymbol{q}$ is very small, the information on lattice strains accessible via GID is restricted to the lateral components of the strain tensor. The displacement field $\boldsymbol{u}(\boldsymbol{r})$ enters into the diffuse intensity via $\exp(\mathrm{i}\boldsymbol{q} \cdot \boldsymbol{u})$. Therefore, only the *radial* direction, i.e. the intensity distribution along the $\boldsymbol{q}$-vector, is strain-sensitive, whereas the angular directions (i.e. perpendicular to $\boldsymbol{q}$) are solely sensitive to electron density fluctuations such as shape, size, and correlation effects. However, refraction effects modify the intensity distribution along q_z. Therefore, the distorted-wave Born approximation has to be considered for the simulation of diffuse scattering in the GID geometry, at least for out-of-plane scattering.

Comprehensive introductions to GID are given in [27, 28, 29].

3.6.3 Grazing-Incidence Small-Angle X-Ray Scattering

HRXRD and GID record the diffuse intensity in the vicinity of an arbitrary reciprocal-lattice point with $\boldsymbol{h} \neq 0$. This leads to large scattering angles. By contrast, the corresponding scattering angles in the proximity of $\boldsymbol{h} = 0$ are rather small. This case is, thus, referred to as small-angle X-ray scattering [30, 31] (SAXS). Following (2.71), there is obviously no influence of strain, and only electron density fluctuations averaged over the atomistic structure are probed. SAXS has become a well-established technique to study the structure of objects in the spatial range from 10 to 1000 nm. However, the conventional transmission geometry is not suitable for mesosocopic objects grown on a thick substrate. Therefore, a reflection geometry with grazing incidence and exit is chosen, combining SAXS with surface sensitivity. This idea was first applied by Levine et al. [32, 33]. Figure 3.8 shows a schematic view of the scattering geometry.

The extended Ewald sphere in the GISAXS geometry is shown in Fig. 3.9. In the vicinity of $\boldsymbol{h} = 0$, the Laue zones restrict the accessible part of reciprocal space to a small area only. However, the scattering vector is restricted in the x direction only, whereas the q_y and q_z components of the scattering vector are not limited. This feature of GISAXS has far-reaching consequences. If, for example, small length scales are to be measured (e.g. short-scale roughness or positional correlation between closely spaced mesoscopic structures), the necessary range in the q_x direction can be reached only at high values of q_z. This, however, would drastically decrease the scattered intensity. Therefore, *coplanar* GISAXS (Fig. 3.9b) is not suitable for measuring electron density fluctuations on a comparatively small length scale.

A better way of performing the experiment is to investigate the diffuse scattering in the q_y–q_z plane, where there is no such restriction. This is, historically, the classical GISAXS geometry applied by Levine et al. [32, 33]. These authors used a linear PSD aligned horizontally with respect to the surface and recorded the intensity distribution along q_y at a fixed value of q_z (Fig. 3.8b). Alternative detector arrangements are possible: by using a CCD (Fig. 3.8a) or a combination of a PSD and a crystal analyzer (Fig. 3.8c),

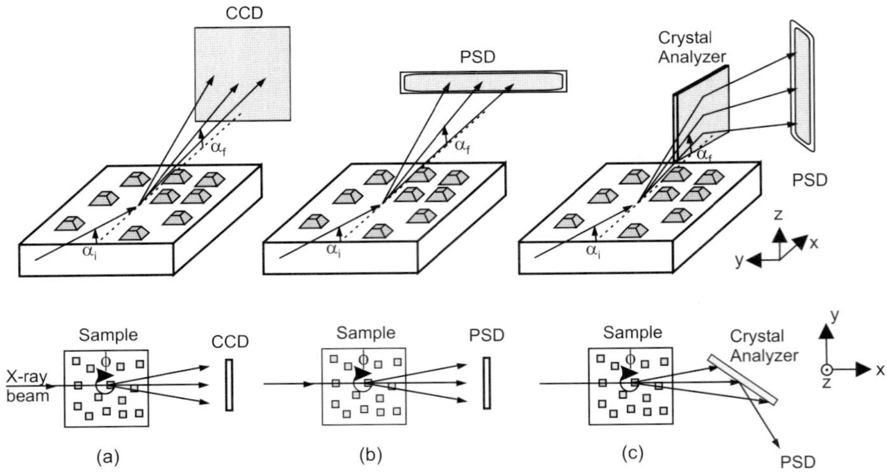

Fig. 3.8. Schematic illustration of the scattering geometry of GISAXS. The incidence angle of the monochromatic X-ray beam, α_i, is set to a value close to the critical angle of total external reflection. The specular intensity ($\alpha_\mathrm{i} = \alpha_\mathrm{f}$) and the diffuse scattering near the specular beam are recorded by either (**a**) a CCD detector, (**b**) a horizontally aligned PSD, or (**c**) a combination of a crystal analyzer and a vertically aligned PSD. By azimuthally rotating the sample, the entire 3D diffuse intensity can be recorded. A selective attenuator can be inserted in order to suppress the specular beam. The intensity of the specular beam is several orders of magnitude higher than the diffuse intensity and thus may cause overflow of the detector

the intensity distribution in the q_y–q_z plane can be collected.[8] Finally, by azimuthally rotating the sample (angle ϕ), the entire three-dimensional intensity distribution of diffuse scattering in the vicinity of $\boldsymbol{h} = 0$ can be recorded without any restriction by the Laue zones. This makes this kind of scanning more appropriate than a measurement along q_x performed by scanning of α_i. Another decisive advantage is that an azimuthal rotation ensures that the angle of incidence α_i – and thus the X-ray penetration depth – is kept constant. As can be seen later in Sect. 3.7, however, the resolution in the q_y direction is one order of magnitude worse than along q_x.

Similarly to GID, a 3D mapping of reciprocal space in the GISAXS geometry requires good collimation in both the α_i and the θ_i directions. When an area detector is utilized (Fig. 3.8a), adequate angular resolution is only achieved for sufficiently small dimensions of the primary beam in the verti-

[8] Strictly speaking, a linear PSD corresponds to a curved line in reciprocal space, and a two-dimensional area detector corresponds to a curved area on the Ewald sphere of the outgoing wave vector. For a sufficiently small aperture of the detector, however, these areas can be approximated as planes, e.g. the q_y–q_z plane.

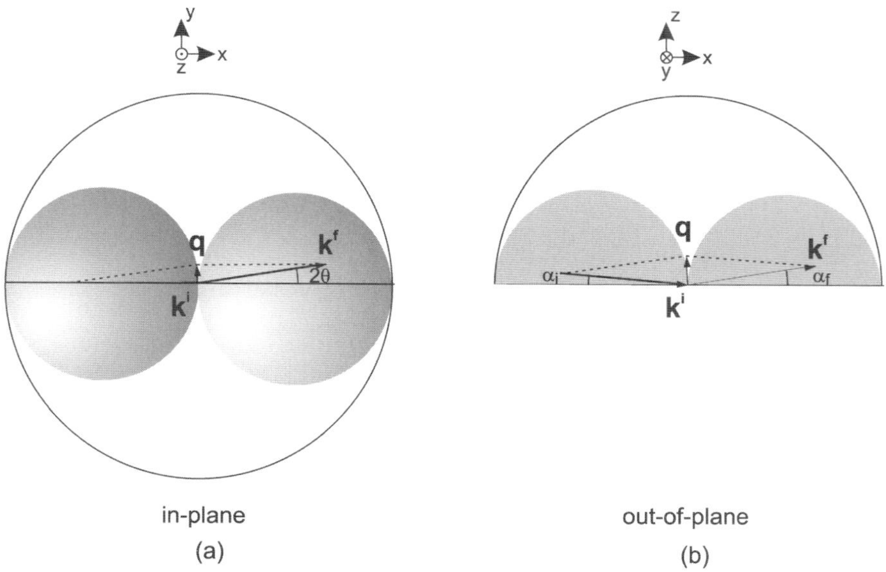

Fig. 3.9. Extended Ewald sphere construction in the GISAXS geometry in an in-plane (**a**) and out-of-plane (**b**) section. The forbidden Laue zones (*shaded gray*) restrict the accessible range for the q_x component of the scattering vector, whereas the q_y and q_z components are not limited

cal and horizontal directions. Larger horizontal beam sizes are possible when a crystal analyzer is used.

The advantage of using a CCD detector is demonstrated in Fig. 3.10, where GISAXS from self-organized SiGe islands grown on Si(001) is presented. As can be seen, the use of a CCD detector yields commensurate results with respect to statistics and resolution; however, the use of a CCD detector (Fig. 3.10c) reduces the data acquisition time by at least a factor of 35 compared with using a combined setup of a PSD and a crystal analyzer (Fig. 3.10b).

Usually, the term GISAXS refers to the investigation of diffuse scattering. Nearly the same setup is used for X-ray specular reflectometry (XSR). However, in XSR the intensity of only the specularly reflected beam is measured.

The simulation of the diffuse intensity in the GISAXS geometry requires the application of the distorted-wave Born approximation. Since refraction and reflection effects (see Figs. 2.13 and 2.14) influence the vertical component of the scattering vector, this requirement holds at least for out-of-plane scattering.

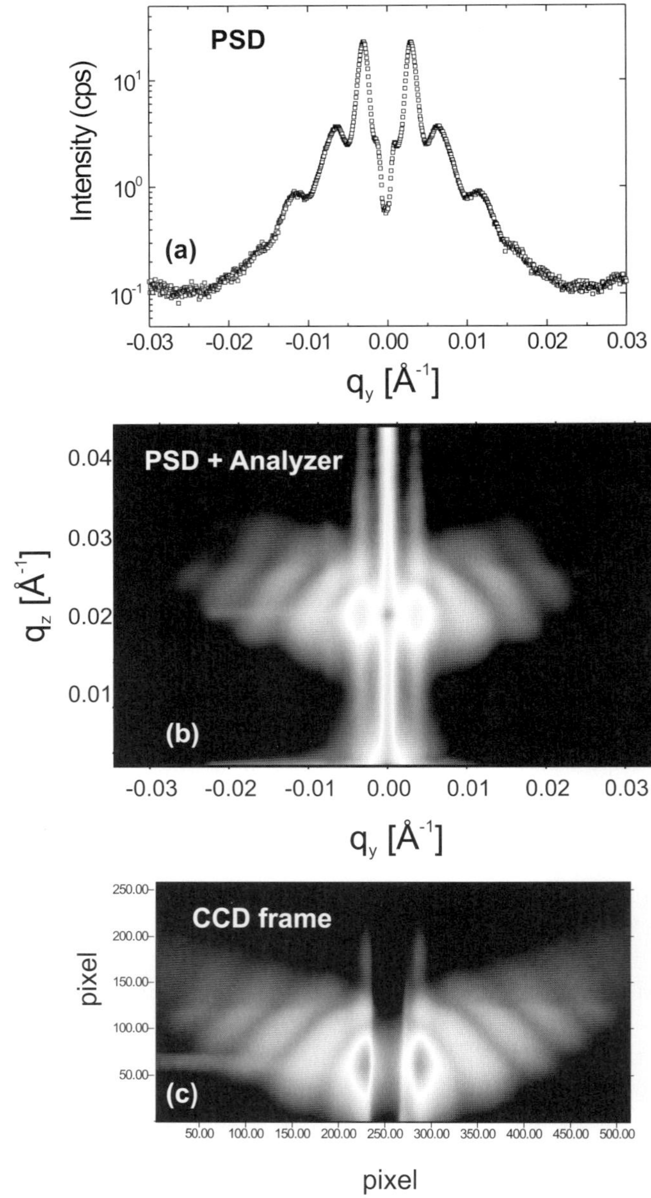

Fig. 3.10. Demonstration of multidetection in the GISAXS scattering geometry. The sample consisted of self-organized free-standing SiGe islands grown on Si(001). (**a**) Horizontal linear PSD (t = 60 min) at $q_z = 0.026\,\text{Å}^{-1}$, (**b**) vertical linear PSD along with a crystal analyzer ($t = 280\,\text{min}$), (**c**) CCD detector ($t = 8\,\text{min}$). The measurements were performed on the same sample at different experimental stations at ESRF. In **(a)** and **(c)** an absorbing wire was used to attenuate the strong specular beam and the 00ℓ truncation rod

3.7 Experimental Resolution

The resolution in reciprocal space is given by

- the collimation of the incident beam,
- the precision of detecting the direction of the scattered wave vector, and
- the energy resolution of the incoming X-ray beam.

High collimation and high energy resolution of the incident wave are achieved through the use of brilliant synchrotron radiation (see Sect. 3.2). Since the vertical collimation of synchrotron radiation is typically one order of magnitude better than that in the horizontal direction, a vertical scattering geometry is often chosen. This is also advantageous with regard to the high horizontal polarization of synchrotron radiation (note the presence of the polarization factor in (2.28)).

The use of synchrotron radiation ensures very good resolution with respect to the incident wave vector $\boldsymbol{k}^{\mathrm{i}}$. Similarly high resolution can be achieved for the scattered wave vector $\boldsymbol{k}^{\mathrm{f}}$ by the use of a crystal analyzer. In that case, however, 3D reciprocal-space mapping is unfeasible ow-

Table 3.1. Typical values for the experimental resolution of various scattering techniques for $\lambda = 1.5$ Å. The values were estimated by assuming a sample–detector distance of 750 mm and a combined detector resolution/X-ray beam size of 100 µm, resulting in $\Delta\alpha_{\mathrm{f}} = 1.3 \times 10^{-4}$ rad. A typical wiggler X-ray source was assumed. For HRXRD (Si 004 reflection) and GID (Si 220 reflection), a vertical scattering geometry was assumed

Technique	$\Delta\theta_{\mathrm{i}}$(rad)	$\Delta\theta_{\mathrm{f}}$(rad)	$\Delta\alpha_{\mathrm{i}}$(rad)	Δq_x (Å^{-1})	Δq_y (Å^{-1})	Δq_z (Å^{-1})
HRXRD[a]	1×10^{-4}	1.3×10^{-4}	2×10^{-5}	3×10^{-4}	8×10^{-4}	5×10^{-4}
HRXRD[b]	1×10^{-3}	1.3×10^{-3}	2×10^{-5}	3×10^{-4}	9×10^{-3}	5×10^{-4}
GID[c]	2×10^{-5}	4×10^{-5}	1×10^{-3}	9×10^{-5}	2×10^{-4}	5×10^{-3}
GID[d]	2×10^{-5}	1.3×10^{-3}	1×10^{-3}	2×10^{-3}	5×10^{-3}	5×10^{-3}
GISAXS[a]	2×10^{-5}	1.3×10^{-4}	1×10^{-4}	1×10^{-5}	2×10^{-4}	9×10^{-4}
GISAXS[c]	2×10^{-5}	4×10^{-5}	1×10^{-4}	1×10^{-5}	2×10^{-4}	9×10^{-4}
GISAXS[e]	2×10^{-5}	1.3×10^{-4}	1×10^{-4}	5×10^{-5}	6×10^{-4}	6×10^{-3}

[a] Using a CCD detector with a spatial resolution of 80 µm.
[b] Using a linear PSD with a spatial resolution of 80 µm and a 1 mm slit.
[c] Using a linear PSD oriented perpendicular to the sample surface along with a crystal analyzer.
[d] Using a linear PSD oriented perpendicular to the sample surface along with a collimation slit system.
[e] Using a linear PSD oriented parallel to the sample surface with a vertical slit ($\Delta\alpha_{\mathrm{f}} = 1.3 \times 10^{-3}$ rad).

ing to the astronomic data collection time. As discussed in the last section, 3D reciprocal-space mapping is best performed by utilizing position-sensitive area detectors. Compared with high-resolution techniques, these area detectors provide medium angular resolution in the range of 10^{-4} rad. Although this resolution is definitely larger than provided by a crystal collimation, it is well adapted to the diffuse scattering from mesoscopic structures.

In Sect. A.1, a general expression for the resolution function is derived. The resolution function depends on the scattering geometry, and expressions for HRXRD, GID, and GISAXS are given. Typical values are presented in Table 3.1.

3.8 Scans in Reciprocal Space

In order to probe reciprocal space by varying the length and direction of the scattering vector $\boldsymbol{q}$, the angles $\alpha_{\mathrm{i,f}}$ and $\theta_{\mathrm{i,f}}$ have to be varied, i.e. the sample and detector have to be rotated. Different scan modes are possible in reciprocal space, and various notations can be found in the literature. Here, a summary of the most important scans that are used in HRXRD, GID, and GISAXS is given.

- *Transverse or α_{i} scan.* In the coplanar GISAXS geometry (i.e. the q_x–q_z scattering plane), this is a simple rocking of the sample by varying α_{i} while the detector is kept fixed. The length of the scattering vector $\boldsymbol{q}$ is constant and $\boldsymbol{q}$ moves on a circle with its center located at $\boldsymbol{h} = 0$. Since only small angles α_{i} are possible, the scattering vector $\boldsymbol{q}$ moves approximately in the *transverse* q_x direction.
- *Specular or longitudinal scan.* The specularly reflected beam fulfills the condition $\alpha_{\mathrm{i}} = \alpha_{\mathrm{f}}$. In the coplanar GISAXS geometry, the sample and detector therefore have to be moved with a 1:2 coupling so that the specular condition is always maintained. This results in a movement in the *longitudinal* q_z direction.
- *Offset scan.* This is practically identical to the longitudinal scan; however, the specular condition is not exactly fulfilled, and α_{f} exhibits a *constant offset* with respect to α_{i}. In this case the diffuse scattering close to the specular beam can be investigated.
- *Detector scan.* This is the counterpart of the transverse scan, but now the detector is scanned while the sample is kept fixed. This corresponds to a variation of α_{f} while α_{i} is constant, or to a scan of θ_{f} while θ_{i} is constant. Both scan directions can be simultaneously achieved by using an area detector (Fig. 3.4).
- *Angular scan.* In the examples mentioned above, the terms *transverse* and *longitudinal* are defined with respect to the (x,y,z) coordinate system, i.e.

with respect to the sample surface. This is appropriate when the morphology of the surface or of interfaces is to be investigated. In contrast, the term *angular scan* is defined to mean all movements where the length of the scattering vector is kept constant. This is valid when the detector is fixed and only the sample is rotated. In this case the scan direction is always perpendicular to the scattering vector. In the HRXRD or GID geometry, this scattering direction is strain-insensitive.

- *Radial scan.* This is the counterpart of the angular scan. Radial scans are measured in the direction along the scattering vector $\boldsymbol{q}$. Since the length of $\boldsymbol{q}$ is varied, a radial scan is strain-sensitive.
- *Out-of-plane scan.* The term *out-of-plane* is mostly used in GID. It corresponds to a detector scan in GISAXS. Owing to the small angles, a detector scan is approximately perpendicular to the sample surface. Therefore, a PSD aligned perpendicularly to the surface is approximately equivalent to a scan in the q_z direction.

All of these scans are schematically shown in Fig. 3.11.

3.9 Critical Comparison of Scattering Techniques

Without doubt, the use of only one of the above-mentioned techniques would seem to be pointless for a complete characterization. For example, GISAXS is sensitive to shape/size and correlation effects only, whereas in the results of HRXRD and GID, strain effects are also included. This mixture of strain, size, shape, and correlation can be disentangled only by investigating the diffuse scattering around *different reciprocal-lattice points*. Therefore, the combined use of HRXRD, GID, and GISXAS is most advantageous and ensures the best reliability. The benefits and drawbacks of the various scattering techniques for investigating the diffuse scattering of mesoscopic structures are summarized in Table 3.2.

Table 3.2. Benefits of various X-ray scattering geometries

Technique	HRXRD	GID	GISAXS
Strain sensitivity	+	+	−
Shape/size/correlation	+	+	++
Interface roughness	−	+	++
Depth selectivity	−	+	+
Multidetection	++	+	++
Theory	Kinematical	DWBA	DWBA

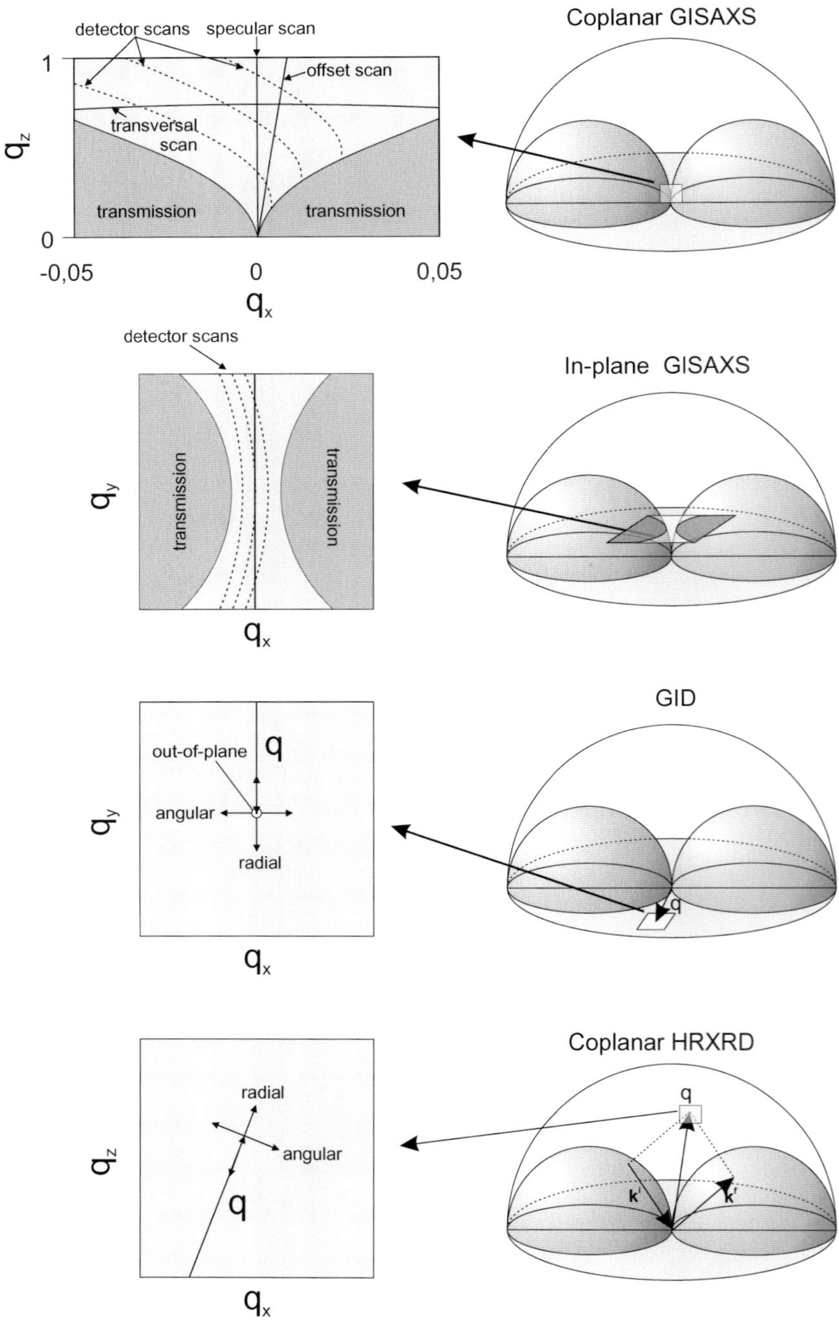

Fig. 3.11. Possible scans in different scattering geometries. For explanation see text

References

1. B. C. Giessen and G. E. Gordon, Science **159**, 973 (1968).
2. B. Buras, J. S. Olsen, L. Gerward, B. Selsmark, and A. Lindegaard-Andersen, Acta Crystallogr. **31**, 327 (1975).
3. H. Metzger, C. Luidl, and U. Pietsch, Nucl. Instrum. Methods A **350**, 398 (1994).
4. C. Kunz, ed., *Synchrotron Radiation* (Springer, Heidelberg, 1979).
5. H. Winick and S. Doniach, eds., *Synchrotron Radiation Research* (Plenum, New York, 1979).
6. E. E. Koch, ed., *Handbook on Synchrotron Radiation*, vol. 1 (North-Holland, Amsterdam, 1983).
7. D. Vaughan, ed., *X-Ray Data Booklet* (Lawrence Berkeley Laboratory, Berkeley, CA, 1986).
8. G. Brown and W. Lavender, *Synchrotron-Radiation Spectra* (North-Holland, Amsterdam, 1991), vol. 3 of *Handbook on Synchrotron Radiation*, chap. 2, p. 47.
9. V. M. Kaganer, B. Jenichen, and K. Ploog, J. Phys. D: Appl. Phys. **34**, 645 (2001).
10. T. Salditt, H. Rhan, T. H. Metzger, J. Peisl, R. Schuster, and J. Kotthaus, Z. Phys. B **96**, 227 (1994).
11. M. Tolan, D. Bahr, J. Süßenbach, W. Press, F. Brinkop, and J. P. Kotthaus, Physica B **198**, 55 (1994).
12. G. Grübel, J. Als-Nielsen, D. Abernathy, G. Vignaud, S. Brauer, G. B. Stephenson, S. G. J. Mochrie, M. Sutton, I. K. Robinson, R. Fleming, et al., ESRF Newslett. **200**, 14 (1994).
13. I. K. Robinson, R. Pindak, R. Fleming, S. B. Dierker, K. Ploog, G. Grübel, D. L. Abernathy, and J. Als-Nielsen, Phys. Rev. B **52**, 9917 (1994).
14. Z. H. Cai, B. Lai, W. B. Yun, I. McNulty, K. G. Huang, and T. P. Russell, Phys. Rev. Lett. **73**, 82 (1993).
15. M. Sutton, S. G. J. Mochrie, T. Grytak, S. E. Nagler, L. E. Berman, G. A. Held, and G. B. Stephenson, Nature **352**, 608 (1991).
16. S. B. Dierker, R. Pindak, R. M. Fleming, I. K. Robinson, and L. Berman, Phys. Rev. Lett. **75**, 449 (1995).
17. B. Lin, M. L. Schlossman, M. Meron, S. M. Williams, Z. Huang, and P. J. Viccaro, Phys. Rev. B **58**, 8025 (1998).
18. E. Jakeman, P. N. Pusey, and J. M. Vaugnan, Opt. Commun. **17**, 306 (1976).
19. J. W. Goodman, *Laser Speckle and Related Phenomena* (Springer, Berlin, 1984).
20. S. K. Sinha, M. Tolan, and A. Gibeaud, Phys. Rev. B **57**, 2740 (1998).
21. A. M. Mathieson, Acta Crystallogr. **A38**, 378 (1982).
22. S. R. Lee, B. L. Doyle, T. J. Drummond, J. W. Medernach, and R. P. Schneider, *Advances in X-Ray Analysis* (Plenum, New York, 1995), vol. 38, p. 201.
23. A. R. Faruqi and C. C. Bond, Nucl. Instrum. Methods **201**, 125 (1982).
24. M. Sonada, M. Takano, J. Miyahara, and H. Kato, Radiology **148**, 833 (1983).
25. M. J. Howes and D. V. Morgan, eds., *Charge-Coupled Devices and Systems* (Wiley, New York, 1979).
26. M. Schuster and H. Göbel, J. Phys. D: Appl. Phys. **28**, A270 (1995).
27. H. Dosch, *Critical Phenomena at Surfaces and Interfaces*, vol. 126 of Springer Tracts in Modern Physics (Springer, Berlin, Heidelberg, 1992).

28. R. Feidenhans'l, Surf. Sci. Rep. **10**, 105 (1989).
29. V. Holý, U. Pietsch, and T. Baumbach, *High Resolution X-Ray Scattering from Thin Films and Multilayers*, vol. 149 of Springer Tracts in Modern Physics (Springer, Berlin, Heidelberg, 1999).
30. A. Guinier and G. Fournet, *Small Angle Scattering of X-rays* (Wiley, New York, 1955).
31. O. Glatter and O. Kratky, eds., *Small Angle X-ray Scattering* (Academic Press, New York, 1982).
32. J. R. Levine, J. B. Cohen, Y. W. Chung, and P. Georgopoulos, J. Appl. Crystallogr. **22**, 528 (1989).
33. J. R. Levine, J. B. Cohen, and Y. W. Chung, Surf. Sci. **248**, 215 (1991).

4 A Model System: LPE SiGe/Si(001) Islands

This section deals with an ideal model system of $Si_{1-x}Ge_x$ islands grown by liquid-phase epitaxy (LPE). These islands are too large to show quantum size effects. However, they are grown rather close to thermodynamic equilibrium and are very regular in shape, size, and chemical composition. Therefore, LPE-grown islands may serve as a model system with regard to a better understanding of the Stranski–Krastanow growth mode. Consequently, we shall first focus on the basic aspects of growth. Beyond the growth aspect, LPE-grown SiGe islands also represent a model system for X-ray diffuse scattering. They are large enough to give sufficiently high scattering signals. On the other hand, the very regular shape and size can be exploited to determine growth-induced composition gradients inside the islands. Besides that, positional correlation effects can be studied and a rich variety of details can be revealed.

4.1 A Brief Introduction to Stranski–Krastanow Growth

The heteroepitaxial growth of a strained layer on top of a substrate leads to an accumulation of elastic energy inside the strained layer which depends on the layer thickness, the lattice mismatch between the layer and substrate materials, and the elastic constants of the layer material. The accumulated energy can be relieved in different ways. One possibility is the formation of misfit dislocations at the interface between the layer and the substrate, inducing *plastic relaxation* inside the layer. From an energetic point of view, this leads to the concept of an equilibrium critical thickness h_c. For layer thicknesses below h_c, the strained state is the thermodynamic equilibrium state and the epitaxial growth of the layer is perfectly coherent with respect to the underlying substrate. For layer thicknesses above h_c, a strained layer may be metastable or it may relax through the creation of misfit dislocations. For semiconductor strained layers, the theory developed by Matthews and Blakeslee [1] for the calculation of h_c is most generally used, although several improved models have been suggested [2, 3, 4, 5, 6, 7]. These models are based on energetic considerations only; that is, the kinetics of misfit dislocation formation – which needs thermal activation – is not considered. Therefore, by choosing appropriate growth conditions, dislocation-free metastable layers can be grown with layer thicknesses far beyond the critical thickness [8].

Besides the generation of misfit dislocations, purely *elastic relaxation* processes may also release the elastic energy. In recent years, the Stranski–Krastanow growth mode [9] has attracted considerable interest. At low coverage, planar (two-dimensional) growth takes place; however, when the coverage is increased above a critical thickness, growth proceeds further through the formation of three-dimensional islands (Fig. 4.1). This may be energetically favored if the gain in elastic energy E^{elastic} is larger than the necessary amount of surface free energy E^{surface}. The total gain in energy can be expressed as a function of the island volume V,

$$E = E^{\text{elastic}} + E^{\text{surface}} = -c_1 V + c_2 V^{2/3} , \tag{4.1}$$

and is sketched in Fig. 4.2.

Equation (4.1) holds for a given island shape; conversely, that shape influences the constants c_1 and c_2. Consequently, the island equilibrium shape depends critically on the balance between surface free energy and elastic energy.

Fig. 4.1. Schematic view of a strained planar layer (*left*) and a coherently grown relaxed island grown on top of a wetting layer (*right*). The island has a larger surface free energy, however, it may relax elastic strain energy.

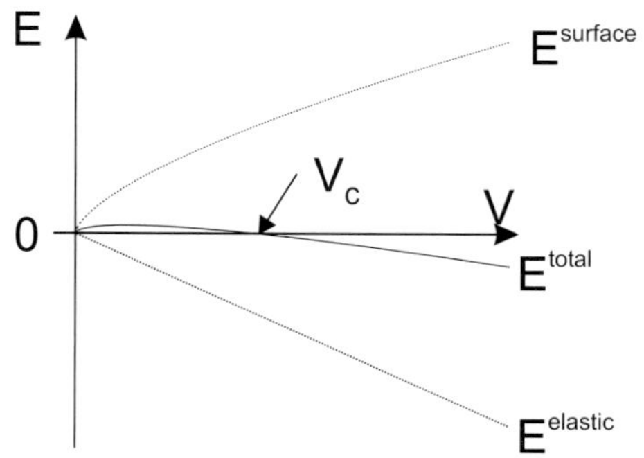

Fig. 4.2. Surface free energy ($E^{\text{surface}} = c_2 V^{2/3}$) and elastic energy of relaxation ($E^{\text{elastic}} = -c_1 V$) of a three-dimensional island as a function of its volume V. The sum ($E^{\text{total}} = E^{\text{surface}} + E^{\text{elastic}}$) represents the total gain in energy as a result of island growth as compared with planar growth. The exact shape of the islands is reflected in the different prefactors c_1 and c_2

The generation of islands is advantageous only if the gain in energy is negative. Therefore, the formation of a strained wetting layer followed by three-dimensional island growth is most favorable. From Fig. 4.2, it is also clear that the islands have a minimum size V_{c}. However, (4.1) is based on very simple physical considerations, as it does not include the chemical composition profile inside the island. Tersoff [10] has included the free energy due to mixing and the chemical potential of a reservoir and found possible stress-induced segregation. On the other hand, the critical island size and nucleation energy can be lowered by decomposition, leading to a strongly enhanced nucleation energy.

The Stranski–Krastanow growth of three-dimensional islands is still not understood in detail. Among the most serious problems are the initial stages of island formation and the experimental observation of a limited island size. Simple energy considerations do not satisfactorily answer these questions, and growth kinetics have to be taken into account. These growth aspects will be discussed in more detail in the following sections.

4.2 LPE Growth of SiGe Islands

As depicted in Fig. 4.3, $Si_{1-x}Ge_x/Si(001)$ islands grown by LPE [12] are shaped like truncated pyramids with {111} side facets and a (001) top facet. The resulting samples consist of coherent, highly monodisperse islands with a uniform shape. They are particularly perfect when the growth conditions are chosen close to thermodynamic equilibrium.

Another important feature of LPE-grown SiGe islands is the ability to tune the island size. As outlined by Dorsch et al. [13], the island base width w is closely related to the lattice mismatch f between $Si_{1-x}Ge_x$ and Si:

$$w \propto f^{-2} \,. \tag{4.2}$$

Consequently, if Vegard's law [14, 15, 16] is assumed to be valid, the island base width scales inversely with the square of the germanium content x. At $x = 1$, i.e. pure Ge islands, a minimum size of about 30 nm can be achieved, which is, however, still too large to show quantum size effects. However, the characteristic properties of LPE-grown SiGe islands make them suitable as a model system in the following sense:

- Since growth is performed close to thermodynamic equilibrium, some light can be shed on the frequently discussed question of to what extent the Stranski–Krastanow growth mode can be discussed in terms of total energy minimization. Though most of the early theoretical work is based on energy considerations [17, 18, 19, 20], the influence of kinetic effects during growth has turned out to be very crucial. Those kinetic limitations play a central role in a detailed understanding of the island growth and they depend strongly on the growth conditions.

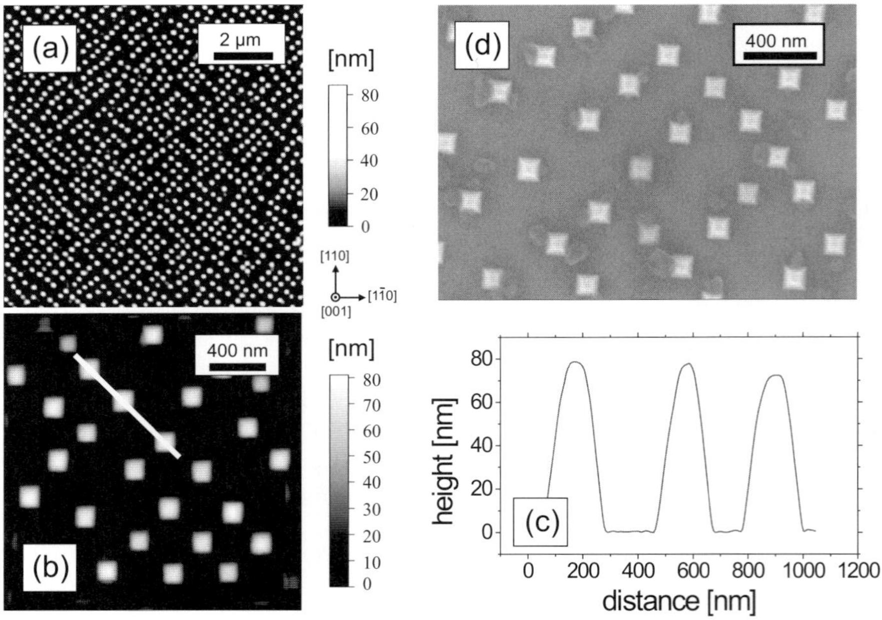

Fig. 4.3. Atomic force (**a**,**b**) and scanning electron (**d**) micrographs of $Si_{0.75}Ge_{0.25}$ nanoscale islands grown on (001) Si by LPE. (**c**) represents an AFM line scan along the *white line* marked in (**b**). Owing to tip convolution, the scan of the top face of the island is rounded instead of showing the exact (001) orientation [11]

- The well-defined size and shape of LPE-grown islands allow one to demonstrate the excellent potential of X-ray diffuse scattering for structural characterization. Moreover, the islands are large enough to provide a sufficiently high scattering signal.

LPE-grown $Si_{1-x}Ge_x$ islands have been the subject of intense work [13, 21, 22, 23, 24, 25]. One critical point is the question of the initial formation of the islands and the origin of the rather perfect alignment of the islands at high island densities. Dorsch et al. [24] attribute this to an underlying ripple pattern of the wetting layer aligned along $\langle 100 \rangle$ that – at later stages of growth – transforms into three-dimensional islands. The corresponding distances between the islands are, therefore, determined by the wavelength of the underlying ripple pattern. Such a pattern is, however, only observed at low ($x < 15\%$) Ge content. Positional correlation is, however, also observed when the ripple pattern is absent ($x > 15\%$), and at low island densities island dimers oriented along $\langle 100 \rangle$ have been found [26, 27]. Therefore, the model suggested by Dorsch et al. needs at least some modifications at high Ge content.

The remainder of the chapter is organized as follows. We first discuss some important phenomenological aspects of the Stranski–Krastanow growth of LPE SiGe islands and, in particular, describe the shape evolution during growth (Sect. 4.3). This is followed by a short discussion of how the island shape and size can be determined by X-ray diffuse scattering (Sect. 4.4). For large islands, i.e. $x < 15\%$, the initial stages of growth can be tracked with real-space methods, e.g. AFM, SEM, or TEM. At higher Ge contents x, however, this is impossible, since the growth of small islands is very fast, so that fully evolved islands are exclusively observed. However, the detection of local chemical-composition changes inside the islands by X-ray diffuse scattering (Sect. 4.5) can be a fingerprint of the existence of different growth stages and, indeed, such a gradient is experimentally found. The evolution of spatial correlation between adjacent islands within the island array is closely related to the initial stages of growth. The general properties of spatial correlation as measured by X-ray diffuse scattering and AFM will be discussed in Sect. 4.7. These results are then compared with kinetic Monte Carlo (KMC) simulations [28].

4.3 Shape Evolution

Dorsch et al. [24] investigated the transition between surface rippling[1] and three-dimensional island growth of SiGe on Si. The elastic surface undulations known as rippling (e.g. [29]) reduce the strain energy of a planar layer at the expense of the surface free energy. Note that surface rippling is an *instability* phenomenon which can be described by elasticity theory [30, 31, 32]. It is therefore of a different nature from three-dimensional island growth, which is a *nucleation* phenomena and thus requires the system to overcome an energy barrier. However, the ripple pattern may serve as a template for further island growth. This can be seen in Fig. 4.4, where both surface rippling and islands are present.

It is striking that the ripple pattern vanishes in regions of high island density (Fig. 4.4b). This depletion of the wetting layer in the immediate environment of the islands has also been observed by other authors [35, 36, 37, 38, 39]. The observations show further (Fig. 4.4a) that the depletion is closely related to the strain energy density ϵ (see also Sect. A.2). The strain energy density is defined as[2]

$$\epsilon = \frac{1}{2} C_{ijkl} \varepsilon_{ij} \varepsilon_{kl} \ , \tag{4.3}$$

where C_{ijkl} are the elements of Hooke's tensor and ε_{ij} are the components of the strain tensor. Around an island, the strain energy density exhibits

[1] *Rippling* is a phenomenon of elastic relaxation of a layer and must not be confused with the *cross-hatch* pattern which appears owing to misfit dislocations at the layer–substrate interface.

[2] Einstein's convention of summation over double identical indices is employed here.

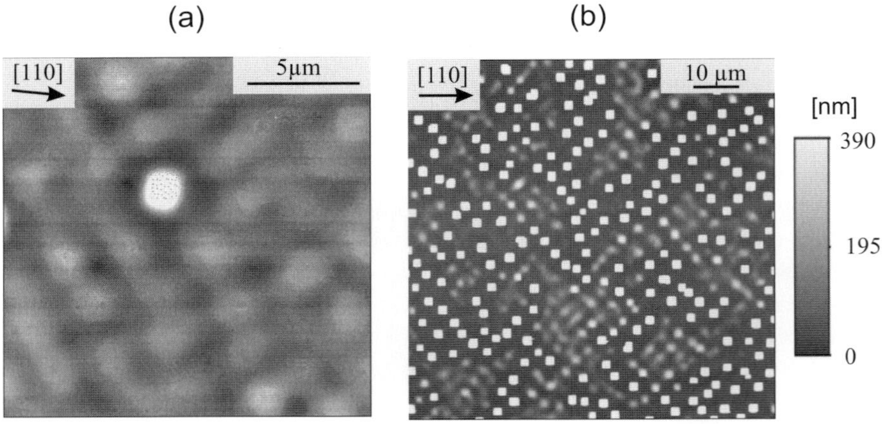

Fig. 4.4. Atomic force micrograph of $Si_{0.9}Ge_{0.1}$ islands for (**a**) low and (**b**) high island coverage. The height scale has been cut off for reasons of better contrast (Hanke [33])

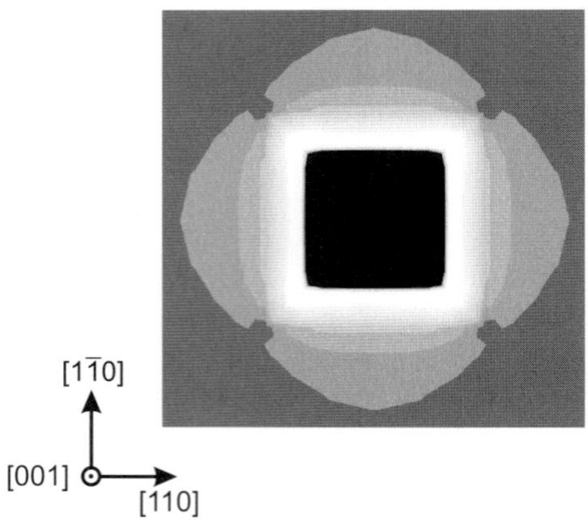

Fig. 4.5. Finite-element numerical calculation of the strain energy density for a truncated pyramid of $Si_{0.75}Ge_{0.25}$. The energy density at the substrate–island interface is displayed [28]

a typical cloverleaf-like shape (Fig. 4.5). The depletion of the wetting layer could be a strong hint of a net material transport from the ripples to the 3D islands. The local growth velocity in LPE is not only proportional to the local oversaturation of the solvent but is also smaller at high elastic stress, i.e. in areas with a high elastic strain energy density. Therefore, the stress may compensate the local oversaturation so that the deposited SiGe in the close environment of the island will be redissolved into the solvent. It should be emphasized that depletion is observed only for rather large islands (as depicted in Fig. 4.4), i.e. at low Ge content ($x < 15\%$), and the size of the

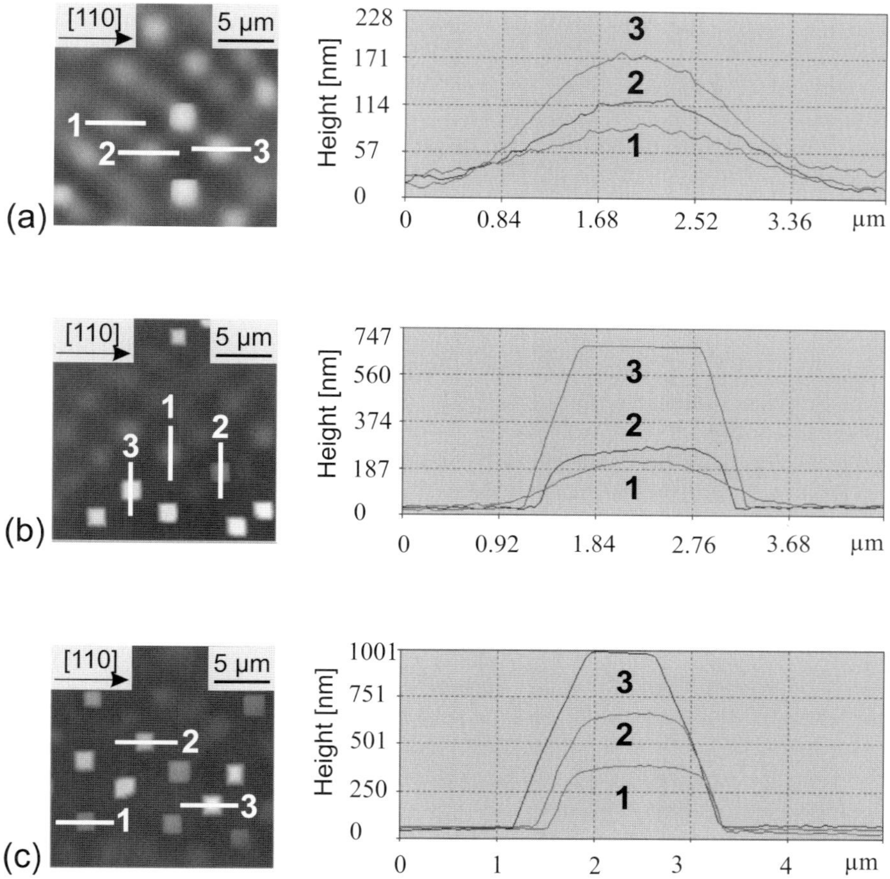

Fig. 4.6. A rich variety of SiGe islands differing in size and shape can be observed by AFM under special growth conditions (for details see text). (**a**) Early stages of growth: constant island base width, evolution of side facets. (**b**) Intermediate stage: at a critical height h_c, {111} side facets have developed. (**c**) Final stages of growth: vertical growth of islands with slightly increasing base widths [34]

depleted region is in the range of a few nanometers. At higher Ge contents ($x > 15\%$), a significant depletion of the wetting layer has not been detected so far.

At low Ge contents it is possible to observe different stages of island growth. This can be achieved by a special growth procedure during LPE. In our application of this procedure, the growth itself was performed at $T = 600°C$ with a gradient of $0.1\,K\,min^{-1}$ and a total growth time of 60 min. Then, the oversaturation was stopped and the sample was kept at a constant temperature of $T = 594°C$ for 18 hours. Under these conditions, a rich variety of islands differing in size and shape can be observed. This is demonstrated in

Fig. 4.6, where atomic force micrographs are presented. In the initial stages of growth (Fig. 4.6a), the islands develop from the ripple pattern and form flat islands with a fixed square base whose size is of the same order as the ripple pattern wavelength. The side faces of these islands exhibit discrete angles, and the largest angles (around 16°) that are observed correspond to {115} facets. At that stage of growth, no top facets have formed as yet. As soon as {115} facets have developed, the islands quickly change to flat truncated pyramids with {111} side facets; however, the island height is still close to the values observed for the {115} islands (Fig. 4.6b). After the {111} side facets have evolved, the growth proceeds mainly in the vertical [001] direction, while the island base width is constant or at least increases only slightly. It may be interesting to mention that an effect of decreasing facet angles that could be called "inverse" has been found [40, 41]. Here, the shape evolution is caused by strain-driven alloying effects.

It is still not clear why the SiGe islands exhibit a *limited, constant size*, i.e. why there is no Ostwald ripening [42, 43, 44, 45, 46].[3] This question does not arise only for SiGe islands but is important for any strained system that is grown in the Stranski–Krastanow growth mode. Kinetic limitations could be responsible for this behavior. For example, recent studies [47, 48] have shown that a kinetic-energy barrier in MBE growth on island facets can lead to self-limiting growth, i.e. the growth rate decreases with increasing island size. It is also very likely that the depletion of the wetting layer plays a central role in these kinetic considerations [49]. However, in LPE the island growth is governed mainly by vertical mass transport from the solute towards the layer and not by surface diffusion. From that point of view, a substantial influence of wetting-layer depletion on growth kinetics is not expected.

Also, energetics might potentially govern the island size distribution. Shchukin et al. [20] have tried to explain the limited size of the islands by taking into account not only the elastic strain relaxation and the surface tension but also the elastic interaction of the edges that are bound the facets of the island. According to this model, there will be a minimum in the energy of the island as a function of the island volume, and a narrow size distribution is favored [50].

4.4 Determination of Shape and Size by GISAXS

In the case of free-standing islands the shape can be evaluated by using AFM or scanning electron microscopy. According to (2.71), however, the shape can also be determined by GISAXS. Free-standing islands are, thus, well suited for comparing the results achieved by direct imaging (AFM and SEM) and

[3] The original meaning of the term *Ostwald ripening* does not imply real growth. Ripening means that a system with a fixed total volume undergoes a transition where all islands coalesce into a single, large island.

by X-ray diffuse scattering. In Fig. 4.3, AFM micrographs of an ensemble of SiGe islands are compared with a scanning electron micrograph recorded at a different position on the same sample. AFM provides information on island shapes and positional correlation. However, the exact island shape information is blurred owing to tip convolution. SEM reveals more detailed information about the island shape, owing to the small size of the scanning electron beam (typically 1–10 nm). However, complete shape information is accessible only if one has both a plan view and a side view of the sample.

The 3D shape of a single island leads to a corresponding 3D intensity distribution in reciprocal space. However, in view of the known symmetry of the islands, two different 2D sections within two nonequivalent mirror planes of the island are sufficient in our case. "2D" means that the scattering vector $\boldsymbol{q}$ maps out a plane in reciprocal space. This is schematically illustrated in Fig. 4.7. Both scattering planes (shaded) are oriented perpendicular to the sample surface and contain either the [110] or the [100] vector.

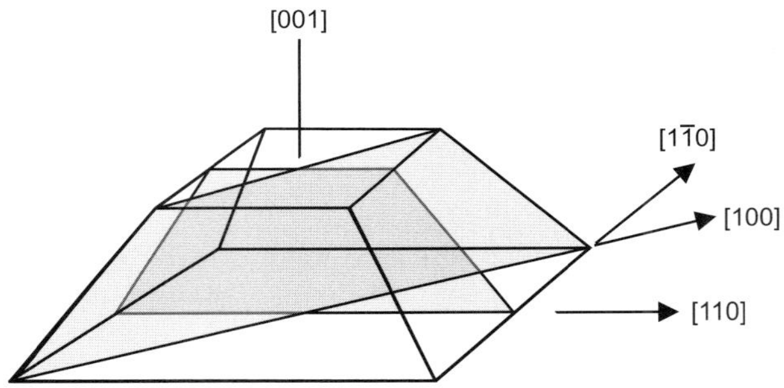

Fig. 4.7. Schematic illustration of island shape and two mirror planes. The orientations of the (*shaded*) mirror planes correspond to the 2D simulation and measurements of diffuse scattering in Figs. 4.8 and 4.9

Let us first neglect dynamical effects and assume that kinematical scattering theory can be used. Then, as quantified in (2.71), the GISAXS signal consists of two main components: the island form factor $|\Omega^{\mathrm{FT}}(\boldsymbol{q})|^2$ arises from the shape and finite size of a single island, whereas the interference function $G(\boldsymbol{q})$ describes positional correlation between different islands. Calculations of the island form factor $|\Omega^{\mathrm{FT}}(\boldsymbol{q})|^2$ for a truncated pyramid, as displayed in Fig. 4.7, are shown in Fig. 4.8. Two significant features will be discussed here:

- There are extended streaks that are collinear with the surface normals of the {111} side facets (as indicated by dashed lines) and with the (001) top facet. In the literature, these streaks are often referred to as *crystal truncation rods* (CTRs), since they appear because of the truncation of a three-dimensional object by a flat surface. The existence of CTRs is quite useful for identifying faceting of mesoscopic structures. Their

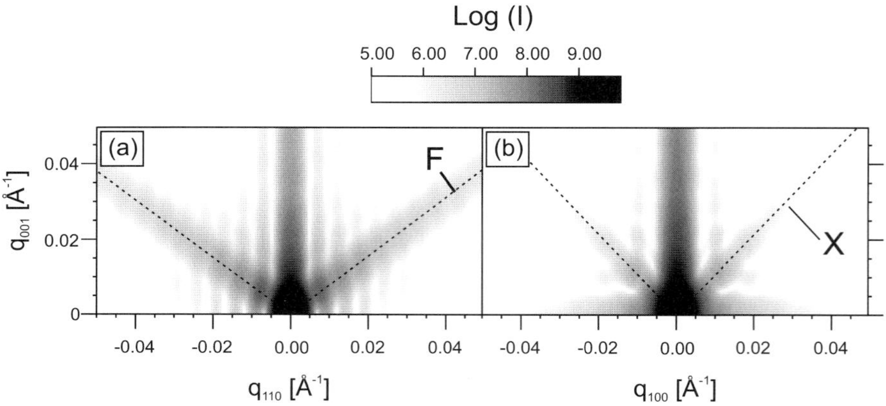

Fig. 4.8. Calculated island form factors $|\Omega^{\mathrm{FT}}(\boldsymbol{q})|^2$ for a truncated pyramid as displayed in Fig. 4.7, for an island base width of $w = 130$ nm and island height of $h = 65$ nm. (**a**) and (**b**) represent different sections through reciprocal space, as indicated in Fig. 4.7. In (**a**), quasi-periodically modulated streaks (marked by *dashed lines*) appear in ⟨111⟩ directions, and are caused by the {111} side facets of the island. The horizontally modulated fringes (F) are caused by the finite horizontal size of the pyramid. In (**b**), a very weak CTR appears at 45° (marked by X), which is caused by the finite with of the island edges [11]

widths scale inversely with the corresponding extension of the facets in real space and thus contain valuable information about the dimensions of the structures.

- In Fig. 4.8a, the intensity distribution is horizontally modulated. These fringes are caused by the finite size of the island. Their spacing in reciprocal space scales inversely with the island size.

The calculated intensity distribution (Fig. 4.8) can be compared with experimental GISAXS measurements (Fig. 4.9). In the experimental data, streaks along the ⟨111⟩ directions appear, thus confirming the existence of {111} side facets. Similarly to the simulations, the intensity along the streaks is modulated. However, there are striking differences between the experiment and the calculations:

- The experimental streaks merge at about $q_z = q_{001} = 0.017$ Å^{-1}, whereas the kinematically simulated streaks merge at $q_x = q_{001} = q_{110} = 0$. Moreover, the {111} streaks are bowed. This behavior is caused by refraction.
- In the kinematical simulations, the specularly reflected beam is missing.
- Close to the specular beam (S), strong correlation satellites (C) appear that show a rodlike intensity distribution.

Obviously, the simple kinematical approach is not able to correctly describe the experimental data. Refraction effects can be included in the theoretical description by using the distorted-wave Born approximation (DWBA)

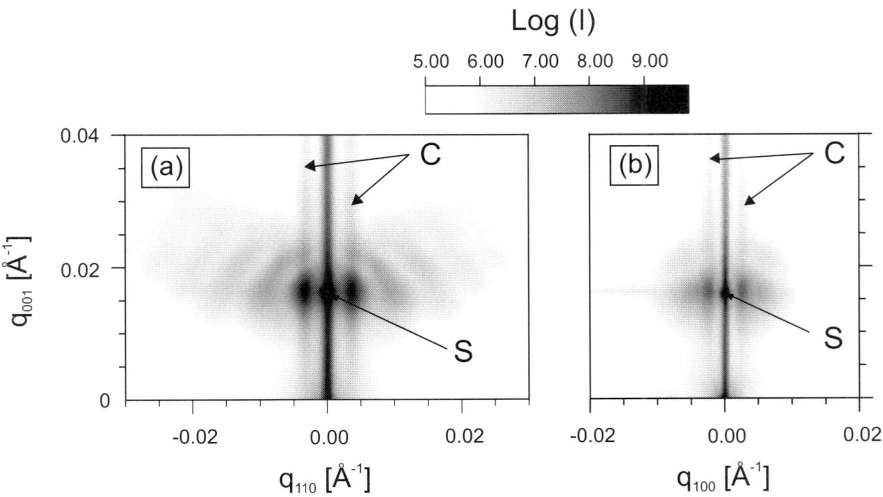

Fig. 4.9. 2D GISAXS measurements of SiGe islands with the in-plane scattering vector (**a**) parallel to [110] and (**b**) parallel to [100]. The plane of (a) contains the surface normals of two {111} side facets. The directions of the streaks along ⟨111⟩ correspond to the surface normals of the corresponding {111} facets. The horizontal intensity fringes along the streaks are caused by the finite size of the islands. Close to the specularly reflected beam (S) at $q_z = 0.016\,\text{Å}^{-1}$, vertical rods (C) show up that are due to positional correlation [11]

introduced in Sect. 2.5. The most serious problem is the theoretical treatment of the specular beam, which is missing in the kinematical theory. As sketched in Fig. 2.14, this beam can also undergo small-angle scattering, leading to enhanced intensity in its close vicinity. The inclusion of the DWBA in the theoretical simulations will be extensively discussed in Chap. 5.

Positional correlation is not a serious problem and will be discussed in more detail in Sect. 4.7. In most cases – when the mean spacing of the mesoscopic structures is larger than their lateral size – the satellite peaks (rods) can be easily distinguished from shape-induced diffuse scattering. Positional correlation can be then described by the function $G(\boldsymbol{q})$ given in (2.53).

4.5 Strain Field and Chemical Composition Profile

For $Si_{1-x}Ge_x$ islands with $x = 25\%$ or larger, no surface rippling can be observed. Here the lattice mismatch between the wetting layer and the Si substrate is larger than 1%, and the corresponding wetting-layer thickness is reduced to a few nanometers only. Therefore, the amplitude of the ripple pattern is expected to be very small (if it exists at all) and is difficult to detect. Furthermore, the growth of a single island is obviously so fast that only fully developed islands are present. However, the detection of a possible chemical composition gradient might serve as a fingerprint of the initial stages of growth. The Ge composition can be visualized by X-ray diffuse scattering. Tiny changes in the Ge concentration x inside a SiGe island lead to a very small effect. Therefore, the main impact on the X-ray diffuse intensity is on the strain field, which is changed by a chemical composition gradient, and not on the structure amplitude, which is, in principle, also sensitive via the different atomic form amplitudes of the different atomic species (see also (2.46)).

The strain distribution of an island of fixed shape, size, and chemical composition (which determines the lattice mismatch through Vegard's law), coherently grown on a substrate, can be calculated numerically by the finite-element method (FEM). This is demonstrated in Fig. 4.10 for the example of a $Si_{1-x}Ge_x$ truncated pyramid with a constant composition of $x = 30\%$. The strain field calculated by the FEM was then used to simulate corresponding X-ray diffuse-scattering intensity distributions (Figs. 4.11 and 4.12). Here, several different Ge concentration profiles were used in order to obtain the best agreement with the experimental data shown in Fig. 4.13.

Let us first focus on the diffuse scattering from a homogeneous SiGe island, i.e. an island with a constant Ge content of $x = 30\%$ (Fig. 4.11a). A more detailed examination of these calculations shows that the origin of the central peak (P1) is scattering from the upper half of the island, whereas the butterfly-shaped diffuse intensity P2 around P1 is caused by diffuse scattering from sections of the island that are close to the island–substrate interface. This behavior is a direct consequence of the spatial strain distribution inside a homogeneous island shown in Fig. 4.10. One of the most important results is that the upper half of the island is strongly relaxed and exhibits cubic lattice symmetry, whereas the lower half of the island is strongly distorted and thus gives rise to broad diffuse scattering. Note that the X-ray simulations were performed in the q_{100}–q_{001} plane. Here, no crystal truncation rods due to the $\{111\}$ side facets are present, which makes the simulations more sensitive to strain-related diffuse scattering. The two main features P1 and P2 are accompanied by weak thickness fringes that are due to the finite island height ($h = 65\,\text{nm}$). The bending of the fringes indicates that they are sensitive to strain. A similar strain-induced deformation of characteristic features in reciprocal space has been found for quantum wire structures [52, 53, 54, 55].

Complete agreement between the simulations of a homogeneous island (Fig. 4.11a) and the corresponding experimental data (Fig. 4.13) is not evident, since the simulated peak positions, as well as the peak widths, shapes, and intensities, are at variance with the experiment. The agreement can be improved by allowing a variation of the Ge composition inside the island, which is the only free fitting parameter as the island size and shape are well known from SEM and AFM. The simulations clearly prove that the Ge content at the top of the island is significantly higher than in regions at the

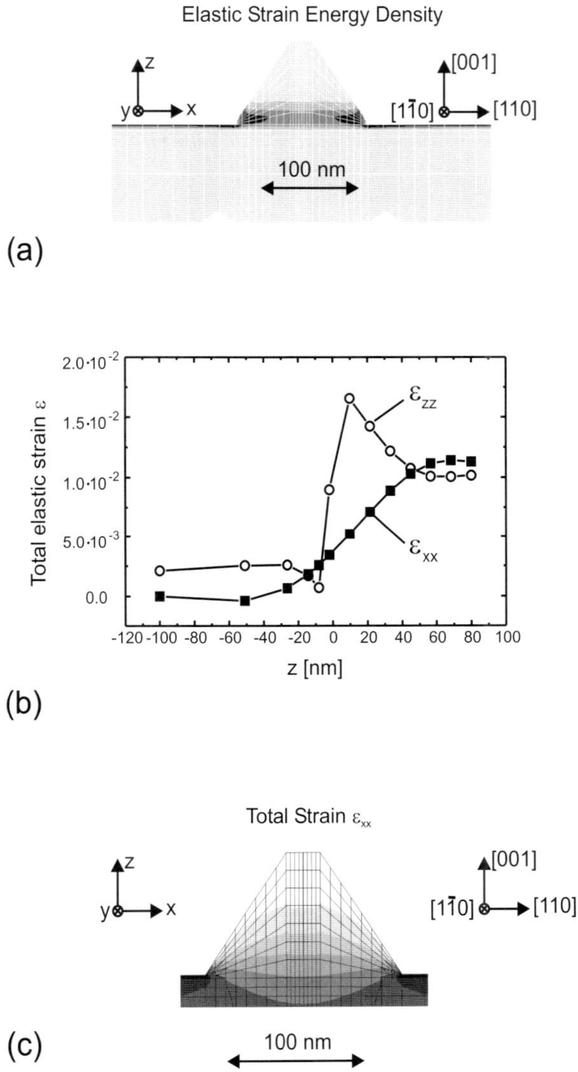

Fig. 4.10. Finite-element calculation for a $Si_{0.7}Ge_{0.3}$ pyramid grown coherently on Si(001). (**a**) Elastic strain energy density ϵ in the (110) plane through the central axis of the pyramid. The contour levels vary from 0 (*bright*) to $2.5 \times 10^7\,\mathrm{J\,m^{-3}}$ (*dark*) on a linear scale. The *thin white lines* represent the FEM mesh used. (**b**) Total strain tensor components $\varepsilon_{zz}\,(z)$ and $\varepsilon_{xx}\,(z)$ along z axis (center of pyramid). $z = 0$ denotes the interface between the wetting layer and the island. (**c**) Total strain tensor component $\varepsilon_{xx}\,(z)$ in the (110) plane through the central axis of the pyramid. The contour levels vary from -4.2×10^{-3} (*dark*) to 1.4×10^{-2} (*bright*) on a linear scale [51]

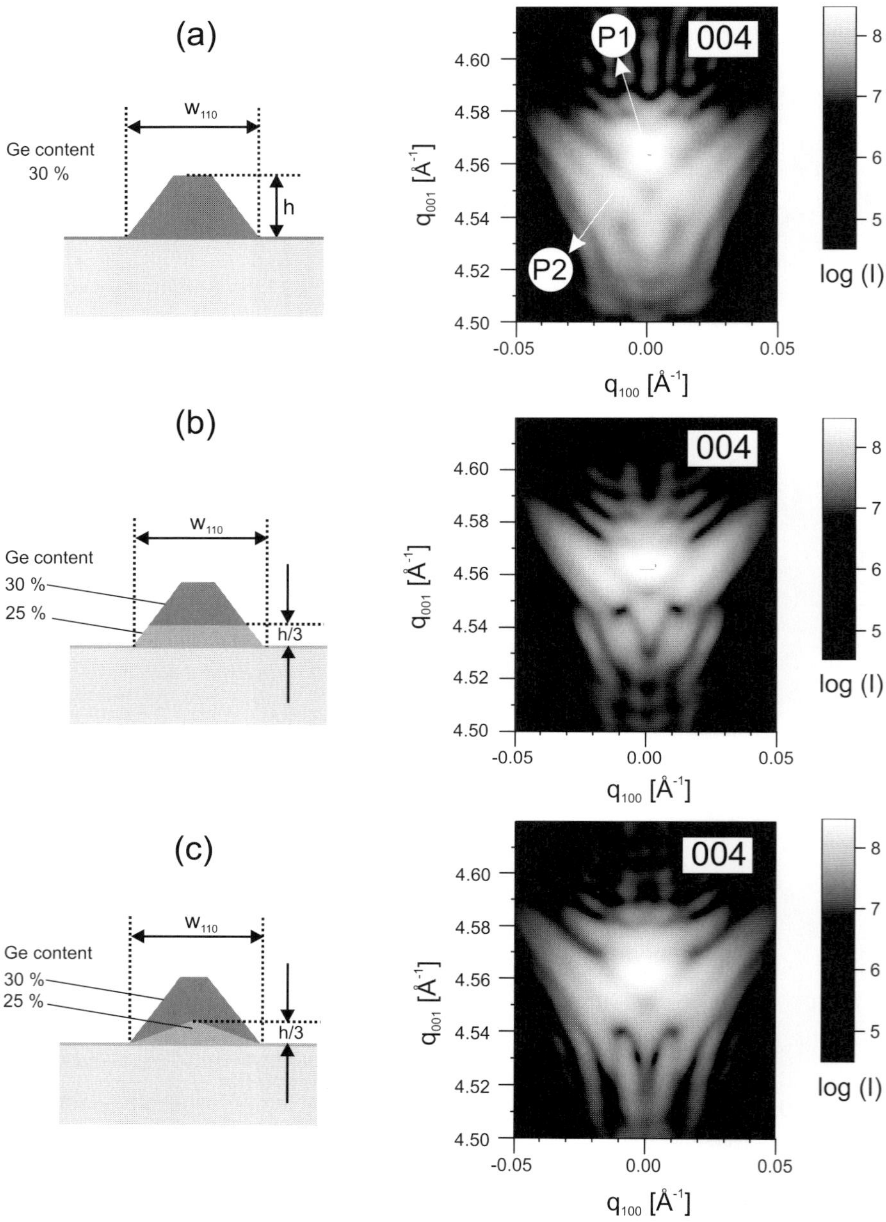

Fig. 4.11. Simulations (*right*) of X-ray diffuse scattering (q_{100}–q_{001} plane) in the vicinity of the symmetrical 004 (HRXRD) reciprocal-lattice point of an $Si_{1-x}Ge_x$ island with an island base width of $w = 130$ nm and height $h = 65$ nm. The corresponding island models, differing in the Ge concentration profile, are shown on the *left*. Scattering from the Si substrate has been omitted

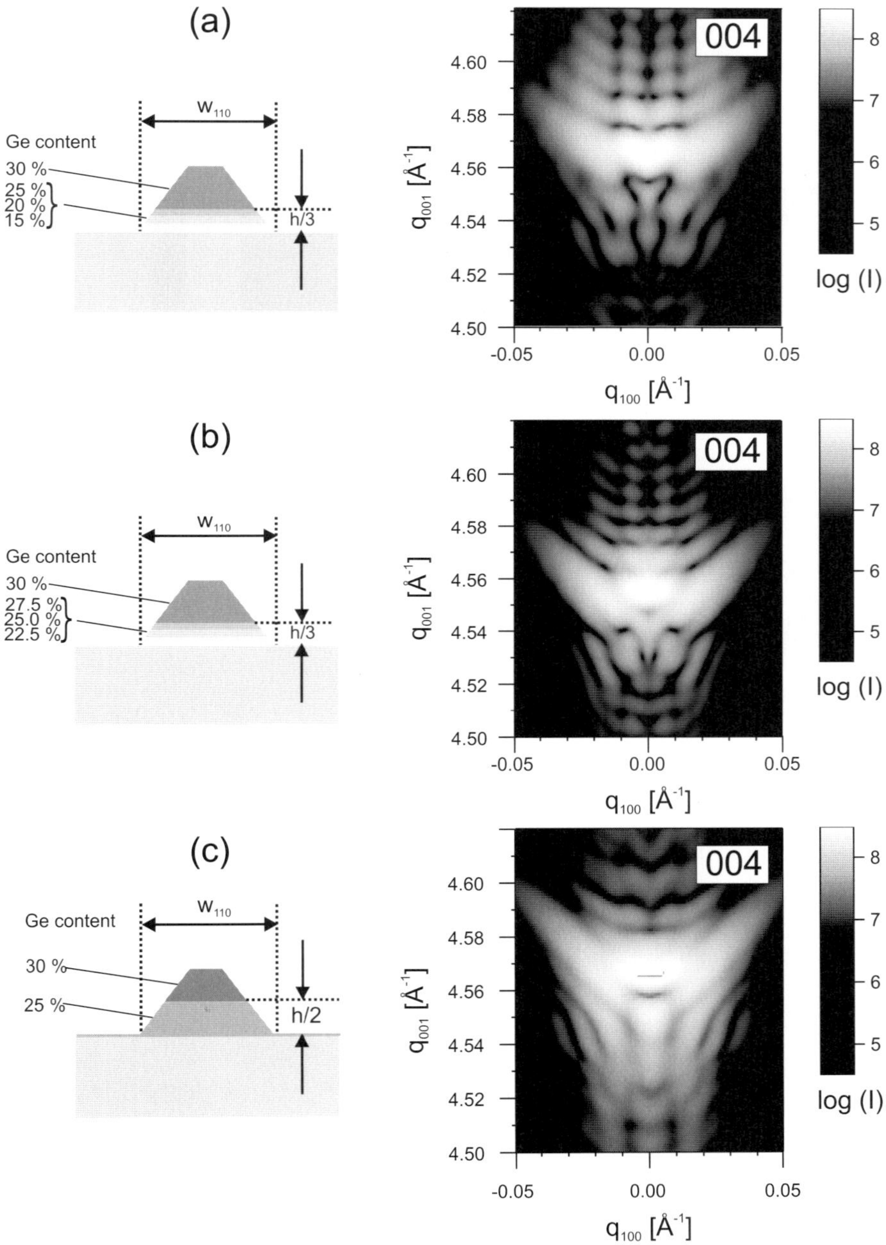

Fig. 4.12. Simulations (*right*) of X-ray diffuse scattering (q_{100}–q_{001} plane) in the vicinity of the symmetrical 004 (HRXRD) reciprocal-lattice point of an $Si_{1-x}Ge_x$ island with an island base width of $w = 130\,nm$ and height $h = 65\,nm$. The corresponding island models, differing in the Ge concentration profile, are shown on the *left*. Scattering from the Si substrate has been omitted

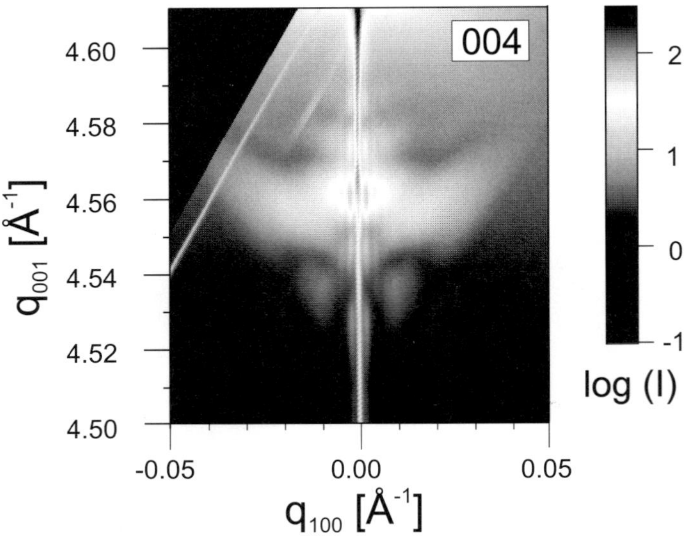

Fig. 4.13. Measured reciprocal-space map (q_{100}–q_{001} plane) in the proximity of the symmetrical 004 reciprocal-lattice point. The Si 004 substrate reflection appears at $q_{001} = 4.628\,\text{Å}^{-1}$ and is not shown here

island base. Surprisingly, an abrupt change of the Ge content at one-third of the island height from $x = 25\%$ to $x = 30\%$ yields the best result (Fig. 4.11b), whereas other simulations using either a linear gradient (Fig. 4.12b), a combination of a linear gradient and a constant Ge content (Fig. 4.12a), or an abrupt change at one-half of the island height (Fig. 4.12c) are not satisfactory. Also, a simulation using an inclusion with shallow {115} facets (Fig. 4.11c) clearly fails.

It is interesting to compare the X-ray diffuse-scattering results with analytical TEM investigations. An EDXS (energy-dispersive X-ray spectroscopy) line profile across the central axis of an SiGe island is shown in Fig. 4.14. Note that the Si and Ge atomic concentrations are not given as absolute values, since they have not been properly normalized to each other (the nominal Ge content is $x = 35\%$). However, the principal features of the EDXS line scan can be discussed. Although a plateau can be identified at the top of the island, a linear gradient of the Ge composition seems to be present at low values of z, rather than an abrupt change as measured by X-ray diffraction. It is also striking that there seems to be a large amount of Ge inside the Si substrate. The discrepancy between the X-ray data and the EDXS results is still not understood. The beam size of the primary electron beam was about 1 nm. The actual spatial resolution is expected to be slightly poorer, since the electron beam was widened owing to the thickness of the sample, of typically 10 nm. It is, however, questionable whether this can explain the discrepancy.

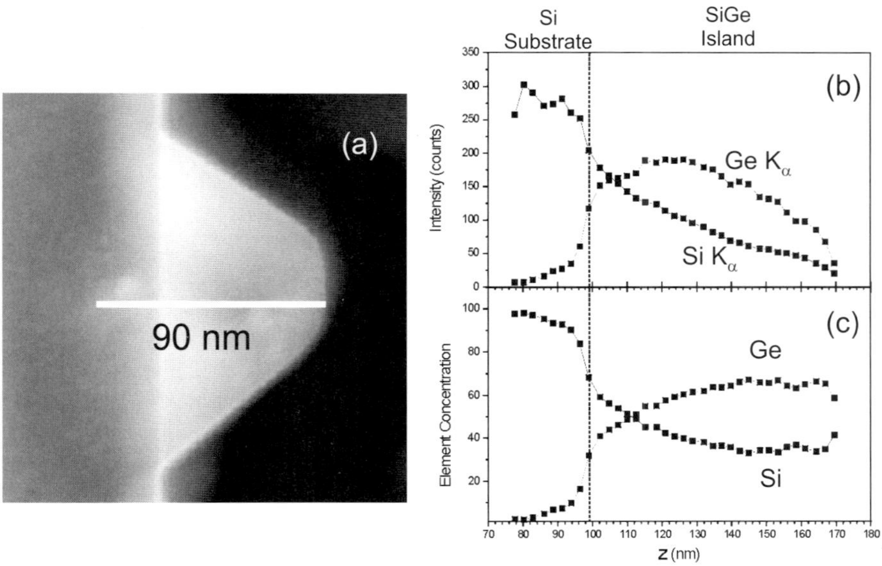

Fig. 4.14. (**a**) EDXS mapping of an Si_xGe_{1-x} island and corresponding line scans: (**b**) Si K_α and Ge K_α fluorescence yield, (**c**) evaluated Si and Ge atomic concentrations. The Si and Ge concentrations are given in arbitrary units. The *vertical dashed line* marks the island–substrate interface. By courtesy of R. Schneider, Humboldt University Berlin

4.6 Geometrical Aspect Ratio

It is interesting to note that the completed coherent (nonrelaxed) $Si_{1-x}Ge_x$ islands have the same shape, independent of the island size: all islands exhibit $\{111\}$ side facets, a (001) top facet, and a vertical aspect ratio of about $Q = w/2h = 1$. At first sight the occurrence of *truncated* pyramids is not self-evident, as one would intuitively expect that a sharp pyramid would release more strain and would, thus, be energetically favored compared with a truncated pyramid. The occurrence of a truncated shape can, however, be understood in the framework of a simple model that will be discussed in the following. The mean total energy gain per atom can be expressed as

$$\Delta E = \frac{A}{N_A \varrho_m} \frac{1}{V} \left(E^{\text{elastic}} + E^{\text{surface}} \right) , \tag{4.4}$$

where A is the molar atomic mass, N_A Avogadro's number, ϱ_m the mass density, and V the island volume. E^{elastic} is the total gain in strain energy and E^{surface} is the total amount of surface free energy. Let us assume, for reasons of simplicity, that the gain in strain energy is proportional to the total volume of the island, while the surface free energy is isotropic. Consequently,

$E^{\text{elastic}} = -c_1 V$ and $E^{\text{surface}} = c_2 S$, where S is the total surface area.[4] The requirement of maximum energy gain ΔE leads to

$$\frac{S}{V} = \text{Min} \; . \tag{4.5}$$

For a pyramid with base width w, top width b, and side angle ϕ (Fig. 4.15), S/V can, after some algebra, be expressed as

$$\frac{S}{V} = 6\frac{w^2 - b^2(1-\cos\phi)}{(w^3 - b^3)\sin\phi} = \frac{6}{w}\frac{1-(1-h/H)^2(1-\cos\phi)}{(1-(1-h/H)^3)\sin\phi} \; . \tag{4.6}$$

A numerical evaluation of (4.5) for truncated pyramids with different angles ϕ is shown in Fig. 4.16 as a function of the reduced height h/H. The minima in S/V are rather shallow; however, their positions suggest a tendency for islands with rather shallow side facets to show higher values of h/H than islands with steep side facets. For {111} side facets, the maximum reduced height is given by $(h/H)_{54.7} = 0.72$, which is equivalent to a vertical aspect ratio of the island of $Q = w/2h = 0.98$ and which fits excellently the experimentally observed values. This surprising result should not be taken too seriously, as it includes neither the spatial distribution of the strain inside the island nor the anisotropy of the surface free energy. Nevertheless, the tendency of islands with shallow side facets to form complete pyramids,

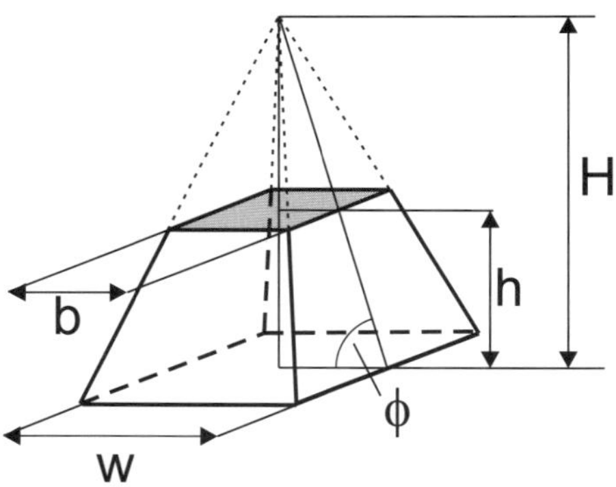

Fig. 4.15. Model of a truncated pyramid with base width w, top width b, height h, and side angle ϕ

[4] The relationship $E^{\text{elastic}} \propto V$ is strictly true only for a homogeneous strain inside the island. However, the FEM calculations indicate that the strain energy density decreases at the top of the island. The relationship $E^{\text{Surface}} \propto S$ is strictly valid for an isotropic surface energy. However, it is well known that faceted surfaces exhibit a smaller surface free energy than vicinal surfaces, and, for example, {111} surfaces are more favorable than {110} and {100} surfaces. Therefore, the two approximations are indeed very rough.

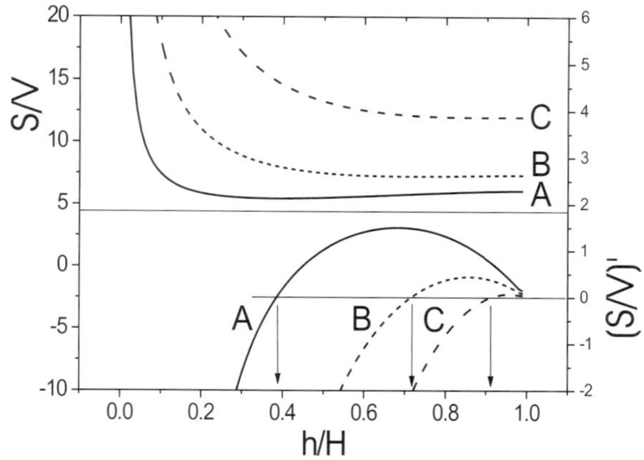

Fig. 4.16. Calculated ratio S/V (*top*) as a function of the reduced height h/H for a truncated pyramid as shown in Fig. 4.15, for different angles ϕ of the side facets: (A) $\phi = 80°$, (B) $\phi = 54.7°$, (C) $\phi = 30°$. (**b**) The positions of the minima can be determined numerically from the first derivative $(S/V)'$ (*bottom*) and are given by $(h/H)_{30} = 0.91$, $(h/H)_{54.7} = 0.72$, and $(h/H)_{80} = 0.39$

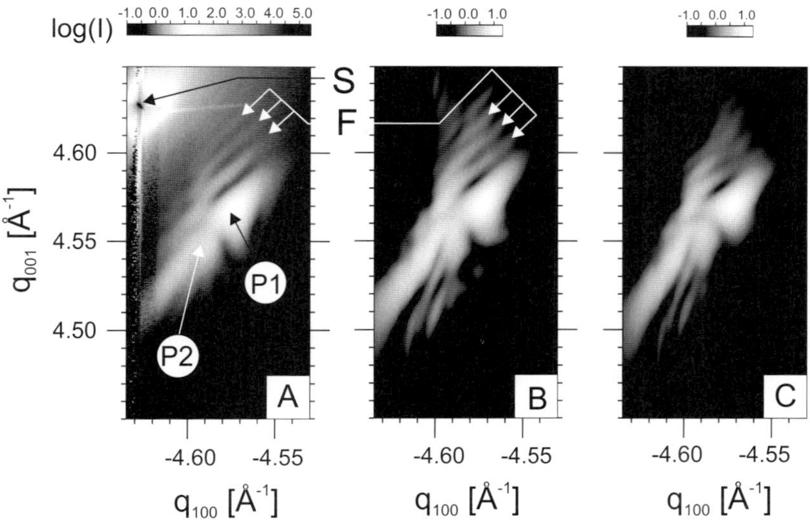

Fig. 4.17. (**a**) Measured X-ray diffuse scattering in the vicinity of the asymmetrical 404 (HRXRD) reciprocal-lattice point of $Si_{1-x}Ge_x$ islands. Similarly to Figs. 4.11 and 4.13, the diffuse intensity is composed of scattering from the relaxed top part of the island (P1) and scattering from the stressed areas close to the island–substrate interface (P2). The substrate peak is marked by S, and thickness fringes by F. Corresponding simulations (model b in Fig. 4.11) were performed using a fixed island base width of $w = 130$ nm but different islands heights: (**b**) $h = 65$ nm ($Q = 1$), (**c**) $h = 78$ nm ($Q = 0.83$). The simulations do not include scattering from the substrate [33]

whereas islands with steep facets show a truncated pyramidal shape seems to be a significant result.

Experimentally, a geometrical aspect ratio of $Q = w/2h = 1$ has been determined by AFM and SEM. Although AFM probes very accurately the island height, tip convolution certainly affects the measurement of the island base width. This can be seen by comparing corresponding AFM and SEM micrographs in Fig. 4.3. X-ray diffuse scattering is very sensitive to the geometrical aspect ratio Q. First, a different aspect ratio will change the periodicity of the thickness fringes. Secondly, the strain field inside the island is modified when the geometrical aspect ratio is altered. In Fig. 4.17, two simulations are compared with experimental results for diffuse scattering in the vicinity of the asymmetrical 404 reciprocal-lattice point. Besides the strain-related features marked by P1 and P2, which are due to scattering from the top and bottom, respectively, of the island, pronounced fringes (F) are present in both experiment and simulations. For $Q = 1$, good agreement between the simulation (Fig. 4.17b) and the experiment (Fig. 4.17a) is evident, while the simulated fringes for $Q = 0.83$ (Fig. 4.17c) are washed out, leading to a significant deterioration of the simulation. In contrast, the strain-related features P1 and P2 seem not to be very sensitive to Q.

4.7 Island–Island Correlation

4.7.1 Evolution of Island Rows

As pointed out in Sect. 4.3, the observation of different growth stages is only possible for a Ge content smaller than $x = 15\%$. At these small Ge concentrations, a periodic ripple pattern exists, which, at later stages of growth, transforms into regular, well-ordered islands. However, as can be seen by SEM and AFM (Fig. 4.3), very regular arrays of SiGe islands can be observed at high Ge concentration also, although no surface rippling can be detected.

A closer inspection of the surface morphology reveals extended island chains (rows) oriented along the $\langle 100 \rangle$ directions. It is very interesting to track the evolution of these island rows as a function of island coverage. At very low island density, there is already a large fraction of islands that are clustered. These clusters consist mainly of island dimers, with a small fraction of linear three-island trimers that are oriented along the $\langle 100 \rangle$ directions (Fig. 4.18a). When the island coverage is increased (Fig. 4.18b–d), the dimers and trimers develop into extended rows of islands aligned along $\langle 100 \rangle$.

The observed behavior clearly proves that the island rows are independently formed: there are no crossing points between the rows, and each island can be unambiguously assigned to a single row. It is also striking that the mean distance between the islands within a single row is very well defined, while there is a significant fluctuation of the distances between neighboring rows. A statistical evaluation of the observed length of the island rows is

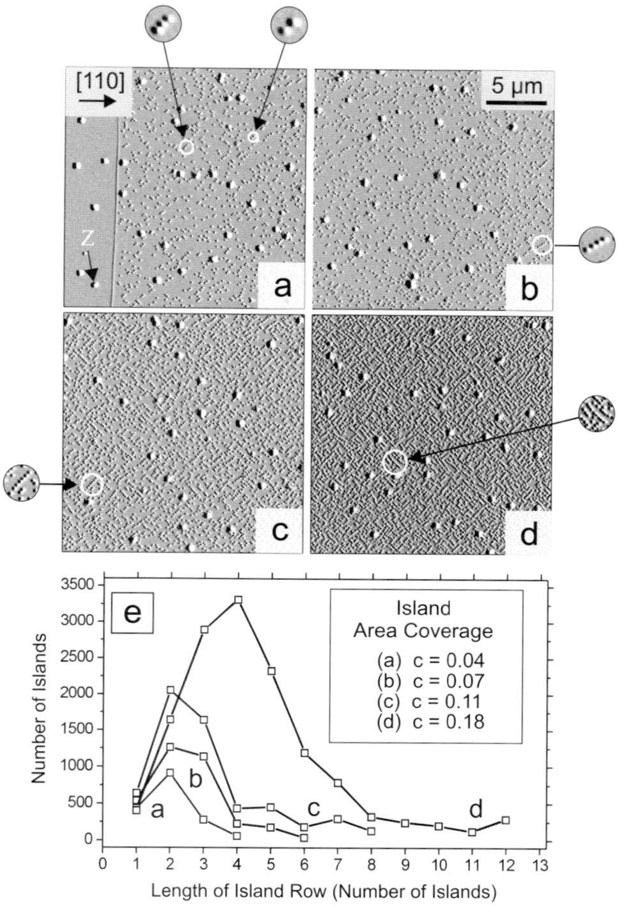

Fig. 4.18. Evolution of spatial ordering of self-organized $Si_{0.70}Ge_{0.30}$ islands ($w = 130\,\mathrm{nm}$, $h = 65\,\mathrm{nm}$) probed by AFM. The growth times were varied between different areas of the same sample, resulting in different island area coverages of (**a**) $c = 0.04$, (**b**) $c = 0.07$, (**c**) $c = 0.11$, and (**d**) $c = 0.18$. (**e**) Statistical distribution of the island rows. After growth, the sample was unfortunately contaminated with Fe silicide particles (Z) [56]

presented in Fig. 4.18e. Three important conclusions can be drawn from the observed behavior:

1. The number of single islands normalized to the total number of islands decreases with increasing coverage.
2. With higher island coverage, the maximum length of the rows increases.
3. The mean length of the rows increases with higher island coverage.

The observed clustering is a clear indication of short-range ordering and suggests that during the growth phase, the nucleation of an island is strongly influenced by already existing islands. One possible explanation is that the extended strain field related to a single island induces additional centers of nucleation, which are preferentially located along the elastically soft direction of the island base diagonal, whereas along the hard ⟨110⟩ directions no such centers are induced. Indeed, the elastic strain energy density (Fig. 4.5) favors island nucleation along the ⟨100⟩ directions. However, the evolution of clusters cannot be ultimately explained in this way, since the elastic strain energy density around a single island does not exhibit any energy *minimum*.

Meixner et al. [28, 57] have shown that the evolution of island dimers and longer rows can be simulated by kinetic Monte Carlo growth simulations. A direct KMC growth simulation is yet not feasible in the case of LPE, where the mass transport is caused by diffusion in the solvent and not – as for MBE – by surface diffusion. Even for MBE, KMC methods cannot treat the three-dimensional island growth of arrays of fully developed islands. However, with appropriately chosen growth parameters and a self-consistently included anisotropic strain field, KMC methods can simulate the initial stages of growth close to equilibrium conditions until platelets of islands arise [58].

Figure 4.19b displays results of a KMC calculation at low island coverage. The simulation was performed close to equilibrium conditions, leading to a qualitative agreement with experimental observations on LPE-grown SiGe islands (Fig. 4.19a) in that island dimers and trimers are also present in the simulations. However, at high island coverage, the experimental island chains

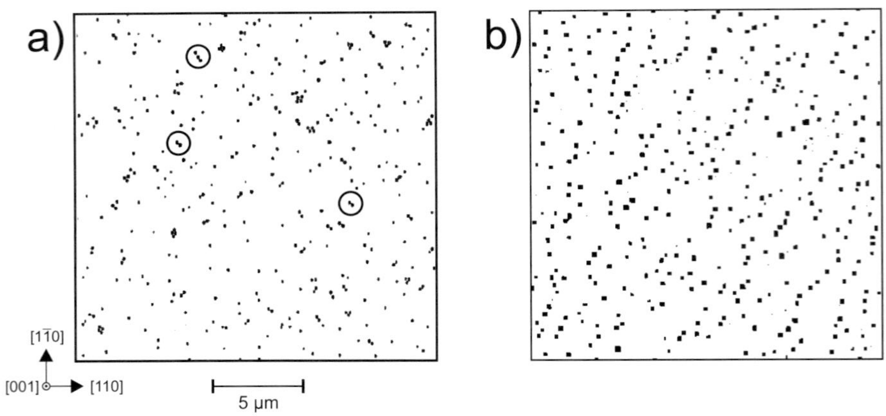

Fig. 4.19. (**a**) Atomic force micrograph of $Si_{0.75}Ge_{0.25}$/Si(001) islands at low island coverage (area coverage $c = 0.02$, excluding the wetting layer). There is a high percentage of islands arranged in dimers and trimers oriented along ⟨100⟩ (some of which are marked by *circles*). (**b**) Results of a kinetic Monte Carlo simulation ($T = 650$ K, growth rate 0.01 monolayers s^{-1}, coverage $c = 0.05$, after 200 s growth interruption) [28]

exhibit a higher degree of ordering. The reason for the poorer spatial ordering in the KMC calculation is the large fluctuations present for any finite simulation time. The approach to equilibrium is driven by local variations of the strain field. The equilibrium state is reached by small strain-induced differences in the diffusion constant. The closer the system is to the equilibrium state, the smaller the driving forces become and the slower the approach to equilibrium becomes. Thus, a perfect ordering can also be expected for KMC simulations but this would require computational power orders of magnitude larger.

4.7.2 Quantitative Analysis of Island Ordering

Owing to its insensitivity to strain, grazing-incidence small-angle X-ray scattering represents a tool that is particularly well suited for the quantitative analysis of spatial correlation effects. In Figs. 4.20–4.23, AFM micrographs and corresponding GISAXS intensity patterns are displayed for varying island area coverage.

Low Island Coverage. At very low island density (Fig. 4.20), the islands seem to be randomly distributed, despite the occurrence of island dimers and linear trimers. However, a ring-shaped intensity pattern appears in the GISAXS distribution, which indicates that there is a constant, isotropic nearest-neighbor distance. The ensemble of islands behaves like a two-dimensional fluid or glass [59] rather than a two-dimensional gas. Eval-

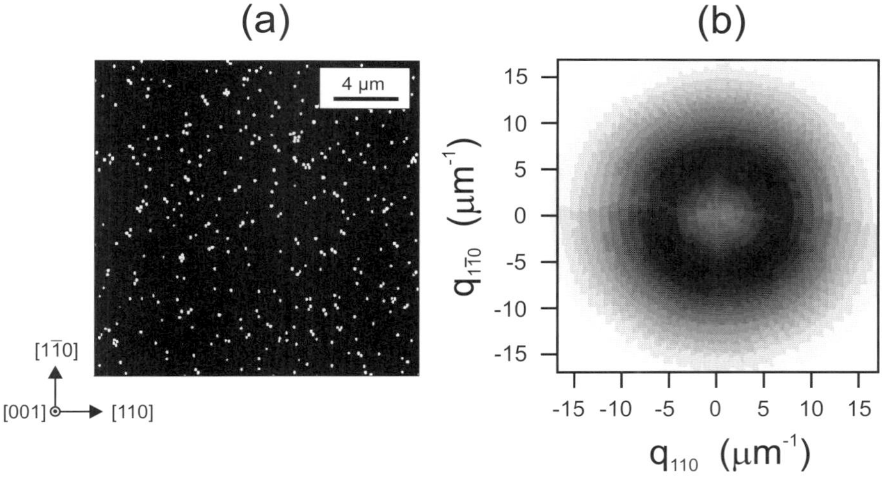

Fig. 4.20. (**a**) AFM micrograph of a sample with low island area coverage ($c = 0.02$, $w = 150\,\mathrm{nm}$, $h = 75\,\mathrm{nm}$), (**b**) corresponding GISAXS in-plane intensity distribution. The grayscale varies varies on a linear scale from *bright* (low intensity) to *dark* (high intensity) [27]

uation of the peak spacings using (2.48) leads to a mean island–island distance of about $\langle d \rangle = 1\,\mu\text{m}$, which is definitely larger than the interisland distance within a dimer. Consequently, GISAXS is sensitive to the mean distance between *clusters* rather than to the distances between the islands within a dimer.

Medium Island Coverage. When the island coverage is increased, the island dimers and trimers develop into chains (rows) oriented along $\langle 100 \rangle$. However, at sufficiently low coverage, the island chains themselves are short and are still not correlated with each other, but are independent. Therefore, each island can be unambiguously assigned to a single chain. This can be seen in Fig. 4.21. The coresponding diffuse intensity now exhibits correlation peaks along the $\langle 100 \rangle$ directions; however, the correlation peaks are smeared out along the $\langle 100 \rangle$ directions. This can be understood if we take account of the coexistence of two characteristic length scales: the island spacing within a row seems to be well defined ($\langle d \rangle = 195\,\text{nm}$), whereas the mean chain–chain spacing is definitely larger and shows a rather broad statistical distribution ($\langle d \rangle = 250\,\text{nm}$–$500\,\text{nm}$). Surprisingly, although they are very weak, correlation peaks along the $\langle 110 \rangle$ directions are also observed in the intensity pattern.

At even higher island coverage (Fig. 4.22), the island rows have developed further. Moreover, missing rows of islands have been filled in so that the distance between the rows now coincides with the distance between the islands within a single row. Therefore, a *face-centered* island pattern has evolved, and the corresponding AFM in-plane power spectrum (Fig. 4.22b) and GISAXS intensity distribution (Fig. 4.22c) show fourfold symmetry with correlation peaks along both the $\langle 100 \rangle$ and the $\langle 110 \rangle$ directions. The corresponding

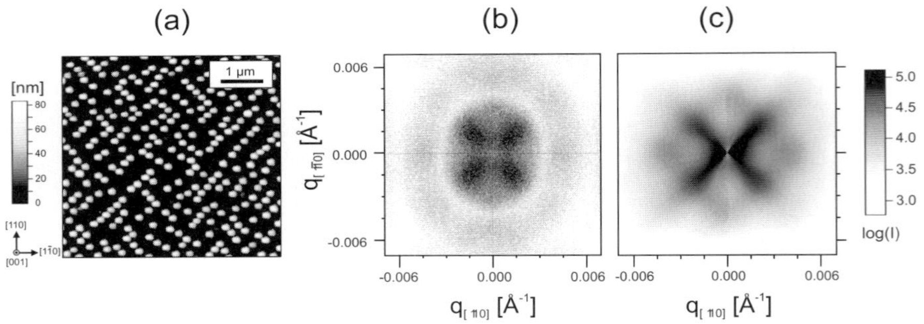

Fig. 4.21. (**a**) AFM micrograph of a sample with medium island area coverage ($c = 0.13$, $w = 110\,\text{nm}$, $h = 55\,\text{nm}$). The islands seem to be elliptically shaped owing to convolution with the AFM tip. (**b**) Corresponding AFM in-plane power spectrum, (**c**) GISAXS in-plane intensity distribution. Correlation peaks are observed in both the $\langle 100 \rangle$ and the $\langle 110 \rangle$ directions

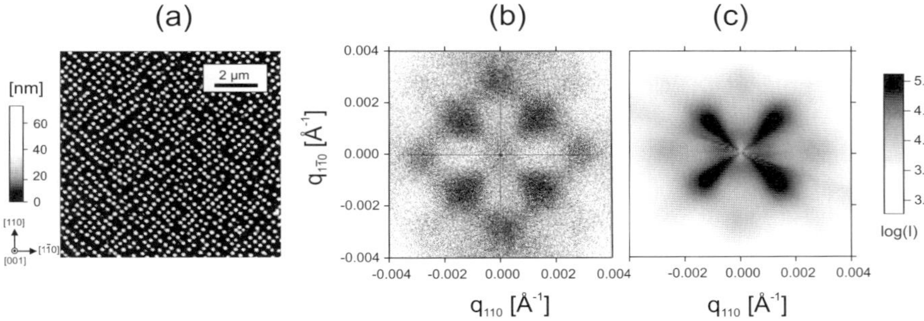

Fig. 4.22. (**a**) AFM micrograph of a sample with medium island area coverage ($c = 0.16$, $w = 130\,\mathrm{nm}$, $h = 65\,\mathrm{nm}$), (**b**) corresponding AFM in-plane power spectrum, (**c**) GISAXS in-plane intensity distribution. Correlation peaks are observed in both the $\langle 100\rangle$ and the $\langle 110\rangle$ directions

mean spacings fulfill the relationship $\langle d_{100}\rangle/\langle d_{110}\rangle = 306\,\mathrm{nm}/224\,\mathrm{nm} \approx \sqrt{2}$, which demonstrates the development of a square island pattern in real space.

High Island Coverage. When the island coverage is further increased, the interisland distances are now so small that the gaps are partly filled by slightly smaller islands (Fig. 4.23). This locally changes the morphology from a face-centered square symmetry to a primitive square symmetry. However, owing to the reduced size homogeneity, the GISAXS intensity pattern does not exhibit exact square symmetry anymore ($(\langle d_{100}\rangle/\langle d_{100}\rangle = 1.21 < \sqrt{2})$, unlike that observed for medium island coverage.

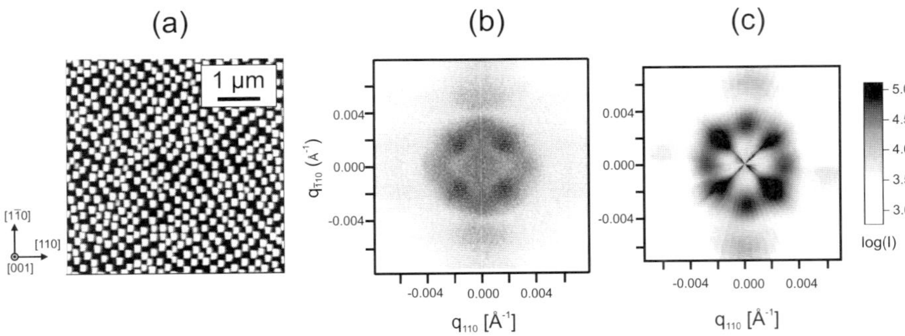

Fig. 4.23. (**a**) AFM micrograph of a sample with high island area coverage ($c = 0.48$, $w = 150\,\mathrm{nm}$, $h = 75\,\mathrm{nm}$), (**b**) corresponding AFM in-plane power spectrum, (**c**) GISAXS in-plane intensity distribution. Correlation peaks are observed in both the $\langle 100\rangle$ and the $\langle 110\rangle$ directions [27]

4.7.3 The Autocorrelation Function in Real Space

The X-ray diffuse scattering data obtained in the GISAXS geometry are in good agreement with the corresponding atomic force micrographs (Figs. 4.20–4.23). However, the island–island correlation is also visible in the X-ray diffuse scattering in the HRXRD geometry, as satellite peaks are clearly resolved in the vicinity of the 00ℓ truncation rod (Fig. 4.13). A section of Fig. 4.13 in the $\langle 100 \rangle$ direction at $q_{001} = 4.560\,\text{Å}^{-1}$ is shown in Fig. 4.24a. It is striking that many orders of correlation peaks are visible, which disperse slightly outwards with increasing horizontal momentum transfer q_{100}. We have pointed out in Sect. 2.3.4 that the occurrence of high-order satellite peaks is not necessarily a consequence of very good spatial correlation, i.e. a large correlation length, but may be generated by an asymmetric line profile in the real-space autocorrelation function $C(r)$. This is indeed the case, as can be seen in Fig. 4.24b. The correlation function $C(r)$ in real space exhibits only the first order peaks, indicating the presence of SRO. However, the asymmetric shape of the first order peaks in $C(r)$ is a strong indication of a *minimum distance* between the islands within a row or, alternatively, a *minimum distance* between the island rows. This is indeed the case, as can be seen in the corresponding AFM micrograph. This behavior is probably caused by short-range repulsion between neighboring islands during growth.

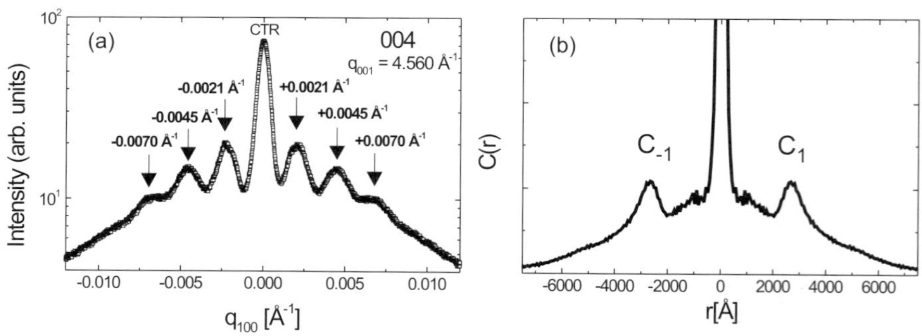

Fig. 4.24. HRXRD from SiGe islands with medium island coverage ($c = 0.20$) in the vicinity of the 004 reciprocal-lattice point. (**a**) Linear section at $q_{001} = 4.560\,\text{Å}^{-1}$ through the experimental intensity distribution shown in Fig. 4.13. The satellite peaks are a consequence of spatial correlation of the islands in the $\langle 100 \rangle$ direction. (**b**) Corresponding autocorrelation function $C(r)$ in real space. The asymmetric profiles of the first order correlation peaks (marked C_{-1} and C_1) in $C(r)$ lead to high-order peaks in the diffuse intensity, which disperse outwards slightly with increasing q_{100}

4.7.4 Inclusion of Spatial Correlation in X-Ray Simulations

The kinematical simulations presented in Figs. 4.11 and 4.12 do not consider spatial correlation. As given by (2.51), the intensity from an array of monodisperse islands can be calculated by multiplying the intensity from a single island by the interference function $G(\boldsymbol{q})$. Since the island shape is well defined, the latter can be obtained, for example by using the GISAXS data discussed above.

Hanke [33] has developed an alternative procedure. Hanke used the present AFM data to construct a corresponding interference function $G(\boldsymbol{q})$. A well-known problem with AFM power spectra is the choice of the area of the surface that is used for the Fourier transformation. If the area is chosen to be similar to the X-ray coherence area (typically a few micrometers square), poor statistics is the main problem. On the other hand, if the area is chosen too large, a pronounced speckle pattern will be present. In a conventional X-ray scattering experiment, the illuminated surface area is much larger than the coherently illuminated area. Consequently, the X-rays average incoherently across the sample surface, with the final result that the speckle pattern is wiped out. The effect of a speckle pattern on the diffuse intensity is demonstrated in Fig. 4.25. Here, the island positions were evaluated from the AFM image and were used to calculate the interference function $G(\boldsymbol{q})$ as defined in (2.52). The total intensity

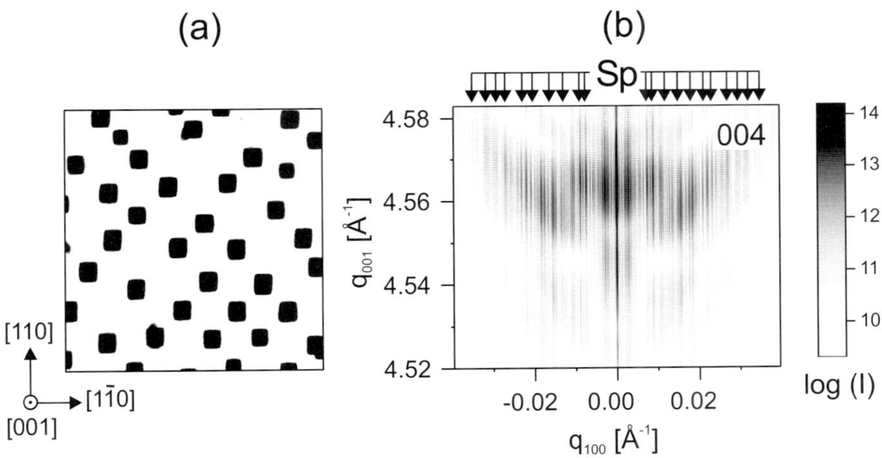

Fig. 4.25. (**a**) Ensemble consisting of 33 SiGe islands on a surface area of 5 µm × 5 µm. (**b**) Simulated intensity distribution in the vicinity of the 004 reciprocal-lattice point. The simulation is based on the structure model given in Fig. 4.11b. Here, the amplitudes of all 33 islands have been coherently summed, leading to a pronounced speckle pattern (marked "Sp") [33]

is thus given by

$$\left(\frac{d\sigma}{d\Omega}\right)_{\text{ensemble}} \propto \left| \boldsymbol{A}^{\text{diffuse}}(\boldsymbol{q}) \sum_{m \epsilon E_n} e^{i\boldsymbol{q}\cdot\boldsymbol{R}_m} \right|^2 = \left| \boldsymbol{A}^{\text{diffuse}}(\boldsymbol{q}) \right|^2 G(\boldsymbol{q}) \,, \tag{4.7}$$

where E_n denotes an ensemble with index n.

From Fig. 4.25b, it is evident that the speckle pattern is present even at high lateral momentum transfer q_{100}. In order to approach the experimental conditions, i.e. to wipe out the speckle pattern, the island positions of the AFM image were modified by allowing random lateral variation of the island positions along the $\langle 100 \rangle$ directions. In this procedure, a variety of new island ensembles (*child ensembles*) were created from the initial *father ensemble*, and the total intensity was then calculated by an incoherent summation (averaging) over all child ensembles:

$$\left(\frac{d\sigma}{d\Omega}\right)_{\text{ensemble}} \propto \sum_{\forall n} \left| \boldsymbol{A}^{\text{diffuse}}(\boldsymbol{q}) \sum_{m \epsilon E_n} e^{i\boldsymbol{q}\cdot\boldsymbol{R}_m} \right|^2 . \tag{4.8}$$

When the maximum lateral island shift is restricted to 20 nm ($\langle d \rangle_{100} =$ 200 nm) and a total number of 1000 child ensembles is used, the speckle pattern is indeed partly wiped out (Fig. 4.26), and correlation peaks (marked C)

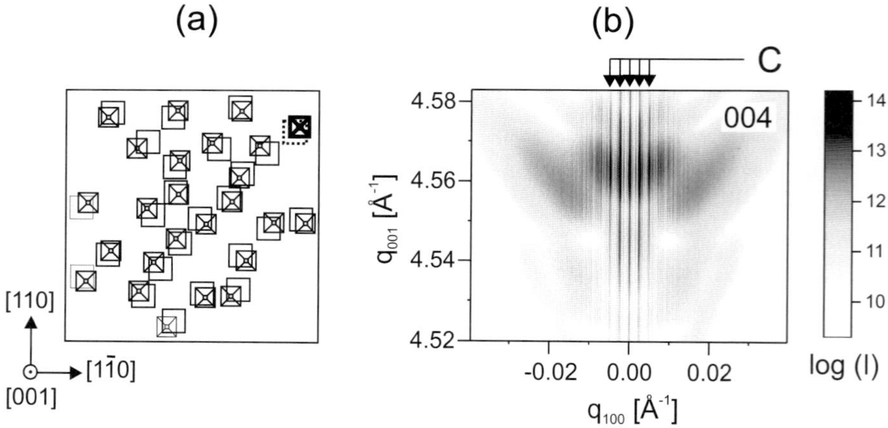

Fig. 4.26. (**a**) Modification of the initial father ensemble consisting of 33 SiGe islands on a surface area of 5 µm × 5 µm. The island positions are shifted randomly along the $\langle 100 \rangle$ directions, creating new child ensembles of islands. (**b**) Corresponding simulated intensity distribution in the vicinity of the 004 reciprocal-lattice point. The simulation was based on the structure model given in Fig. 4.11b. Here, the amplitudes of all 33 islands within the child ensembles were coherently summed. Then the intensities of 1000 child ensembles were added, leading to correlation peaks (denoted by C) [33]

appear that are in qualitative agreement with experiment (Fig. 4.13). However, complete agreement is not evident. Obviously, the degree of disorder inside the father ensemble is not high enough.

Better agreement is achieved when different father ensembles are used. It turns out that the number of islands in such father ensembles need not be high, and instead of using ensembles with areas of 5 μm × 5 μm it is sufficient to treat the problem with ensembles consisting of about 10 islands within an area of 1 μm × 1 μm. This corresponds to the coherently illuminated area in the experiment. When 10 different father ensembles were chosen at different positions on the sample and 1000 child ensembles were created from each father ensemble, very good agreement between experiment and simulation could be achieved. This is demonstrated in Fig. 4.27. Now, the speckle pattern has completely vanished and equidistant correlation peaks (C), exhibiting the correct positions, width, and intensity, are observed. The good agreement can also be seen in Fig. 4.28, where a horizontal section at $q_{001} = 4.560\,\text{Å}^{-1}$ is presented for both experiment and simulation.

In conclusion, it has been demonstrated that AFM data along with X-ray diffuse scattering can be used to extract valuable statistical information about positional correlation. In the simulations, the X-ray coherence length has to be taken into account by restricting the area of the father ensembles to the actual coherently illuminated surface area of the sample. The statistical incoherent averaging over different ensembles of islands can be emulated by

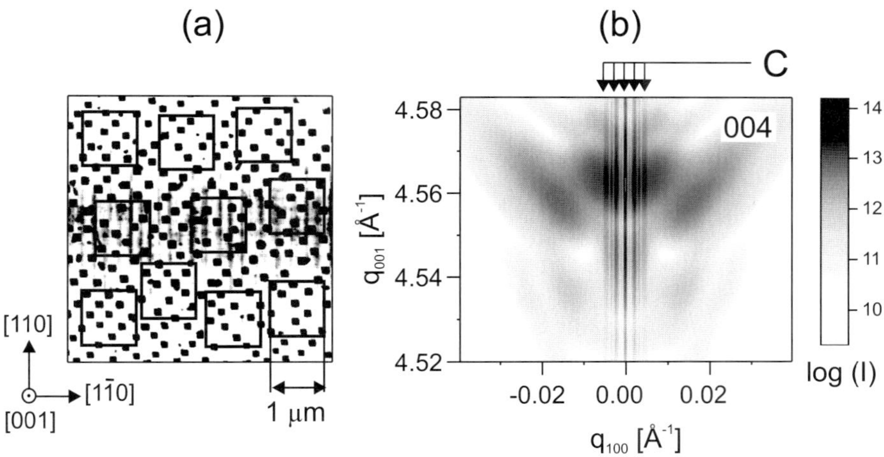

Fig. 4.27. (**a**) Different father ensembles of area 1 μm × 1 μm containing about 10 islands each. From each of these father ensembles 1000 child ensembles were created. (**b**) Corresponding simulated intensity distribution in the vicinity of the 004 reciprocal-lattice point. The simulation was based on the structure model given in Fig. 4.11b. The amplitude of each child ensemble was calculated coherently. Then the intensities of all child ensembles were summed, leading to pronounced correlation peaks (denoted by C) [33]

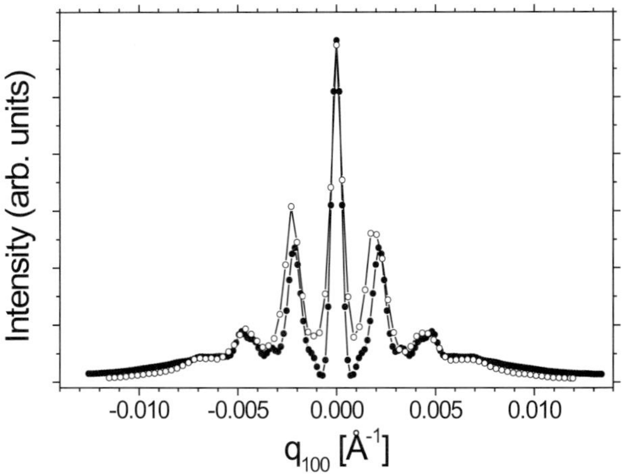

Fig. 4.28. Experimental (∘) and simulated (•) diffuse intensity in the vicinity of the 004 reciprocal-lattice point at $q_{001} = 4.560\,\text{Å}^{-1}$ [33]

allowing random fluctuations of the island positions. However, a single father ensemble obviously produces child ensembles that are too similar to each other for them to be used as a relevant statistical average. Therefore, several different father ensembles have to be used. There are still small differences between experiment and simulation. These can be minimized by further increasing the number of father ensembles.

References

1. J. Matthews and A. Blakeslee, J. Cryst. Growth **27**, 118 (1974).
2. R. People and J. C. Bean, Appl. Phys. Lett. **47**, 322 (1985).
3. R. People and J. C. Bean, Appl. Phys. Lett. **49**, 229 (1986).
4. B. Dodson and J. Tsao, Appl. Phys. Lett. **51**, 1325 (1987).
5. R. Cammarata and K. Sieradzki, Appl. Phys. Lett. **55**, 1197 (1989).
6. D. Chidambarro, G. Srinivasan, B. Cunningham, and C. Murthy, Appl. Phys. Lett. **57**, 1001 (1990).
7. K. Shintani and K. Fujita, J. Appl. Phys. **12**, 7842 (1994).
8. F. Schäffler, Semicond. Sci. Technol. **12**, 1515 (1997).
9. I. Stranski and L. Krastanow, Sitzungsberichte d. Akademie d. Wissenschaften in Wien, Abt. IIb **146**, 797 (1937).
10. J. Tersoff, Phys. Rev. Lett. **81**, 3183 (1998).
11. M. Schmidbauer, M. Hanke, and R. Köhler, Cryst. Res. Technol. **37**, 3 (2002).
12. E. Bauser, *Crystal Growth of Electronic Materials* (Elsevier Science, Amsterdam, 1985), p. 41.
13. W. Dorsch, H. P. Strunk, H. Wawra, G. Wagner, J. Groenen, and R. Carles, Appl. Phys. Lett **72**, 179 (1998).
14. L. Vegard, Z. Phys. **5**, 17 (1921).

15. L. Vegard, Z. Krist. **6**, 239 (1928).
16. A. R. Denton and N. W. Ashcroft, Phys. Rev. A **43**, 3161 (1991).
17. Q. Xie, A. Madhukar, P. Chen, and N. Kobayashi, Phys. Rev. Lett. **75**, 2542 (1995).
18. J. Tersoff, C. Teichert, and M. G. Lagally, Phys. Rev. Lett. **76**, 1675 (1996).
19. V. Holý, G. Springholz, M. Pinczolits, and G. Bauer, Phys. Rev. Lett. **83**, 356 (1999).
20. V. A. Shchukin, N. N. Ledentsov, P. S. Kop'ev, and D. Bimberg, Phys. Rev. Lett. **75**, 2968 (1995).
21. S. Christiansen, M. Albrecht, H. P. Strunk, and H. J. Maier, Appl. Phys. Lett. **64**, 3617 (1994).
22. S. Christiansen, M. Albrecht, H. P. Strunk, P. O. Hansson, and E. Bauser, Appl. Phys. Lett. **66**, 574 (1995).
23. J. Groenen, R. Carles, S. Christiansen, M. Albrecht, W. Dorsch, H. P. Strunk, H. Wawra, and G. Wagner, Appl. Phys. Lett. **71**, 3856 (1997).
24. W. Dorsch, B. Steiner, M. Albrecht, H. P. Strunk, H. Wawra, and G. Wagner, J. Cryst. Growth **183**, 305 (1998).
25. M. Cazayous, J. Groenen, F. Demangeot, R. Sirvin, M. Caumont, T. Remmele, M. Albrecht, S. Christiansen, M. Becker, H. P. Strunk, et al., J. Appl. Phys. **91**, 6772 (2002).
26. M. Schmidbauer, T. Wiebach, H. Raidt, M. Hanke, R. Köhler, and H. Wawra, Phys. Rev. B **58**, 10523 (1998).
27. M. Schmidbauer, T. Wiebach, H. Raidt, M. Hanke, R. Köhler, and H. Wawra, J. Phys. D: Appl. Phys. **32**, 230 (1999).
28. M. Meixner, E. Schöll, M. Schmidbauer, H. Raidt, and R. Köhler, Phys. Rev. B **64**, 245307 (2001).
29. A. G. Cullis, D. J. Robbins, A. J. Pidduck, and D. J. Smith, J. Cryst. Growth **123**, 333 (1992).
30. R. J. Asaro and W. A. Tiller, Metall. Trans. **3**, 1789 (1972).
31. M. A. Grinfel'd, Sov. Phys. Dokl. **31**, 831 (1986).
32. D. J. Srolovitz, Acta Metall. **37**, 621 (1989).
33. M. Hanke (2002), *Streuung von Röntgenstrahlung an selbstorganisierten Halbleiter-Inselstrukturen*, Ph.D. thesis, Humboldt-Universität zu Berlin (Mensch und Buch Verlag, Berlin, 2002).
34. M. Hanke, M. Schmidbauer, D. Grigoriev, H. Raidt, P. Schäfer, R. Köhler, A.-K. Gerlitzke, and H. Wawra, Phys. Rev. B, submitted (2003).
35. U. Denker, O. G. Schmidt, N. Y. Jin-Phillip, and K. Eberl, Appl. Phys. Lett. **78**, 3723 (2001).
36. S. A. Chaparro, Y. Zhang, and J. Drucker, Appl. Phys. Lett **76**, 3534 (2000).
37. X. Z. Liao, J. Zou, X. F. Duan, D. J. H. Cockayne, R. Leon, and C. Lobo, Phys. Rev. B **58**, R4235 (1998).
38. X. Z. Liao, J. Zou, D. J. H. Cockayne, J. Qin, Z. M. Jiang, X. Wang, and R. Leon, Phys. Rev. B **60**, 15605 (1999).
39. X. Z. Liao, J. Zou, D. J. H. Cockayne, Z. M. Jiang, X. Wang, and R. Leon, Appl. Phys. Lett. **77**, 1304 (2000).
40. C. J. Huang, Y. H. Zuo, D. Z. Li, B. W. Cheng, L. P. Luo, J. Z. Yu, and Q. M. Wang, Appl. Phys. Lett. **78**, 3881 (2001).

41. T. I. Kamins, G. Medeiros-Ribeiro, D. A. A. Ohlberg, and R. S. Williams, Appl. Phys. A: Mater. Sci. Process. **67**, 727 (1998).
42. W. Ostwald, Z. Phys. Chem **22**, 289 (1897).
43. W. Ostwald, Z. Phys. Chem **34**, 495 (1900).
44. S. H. Xin, P. D. Wang, A. Yin, C. Kim, M. Dobrowolska, J. L. Merz, and J. K. Furdyna, Appl. Phys. Lett. **69**, 3884 (1996).
45. S. Lee, I. Daruka, C. S. Kim, A.-L. Barabási, J. L. Merz, and J. K. Furdyna, Phys. Rev. Lett. **81**, 3479 (1998).
46. F. M. Ross, J. Tersoff, and R. M. Tromp, Phys. Rev. Lett. **80**, 984 (1998).
47. D. E. Jesson, G. Chen, K. M. Chen, and S. J. Pennycook, Phys. Rev. Lett. **80**, 5156 (1998).
48. M. Kästner and B. Voigtländer, Phys. Rev. Lett. **82**, 2745 (1999).
49. L. G. Wang, P. Kratzer, M. Scheffler, and N. Moll, Phys. Rev. Lett. **82**, 4042 (1999).
50. V. A. Shchukin, N. N. Ledentsov, M. Grundmann, P. S. Kop'ev, and D. Bimberg, Surf. Sci. **352-354**, 117 (1996).
51. T. Wiebach, M. Schmidbauer, M. Hanke, H. Raidt, R. Köhler, and H. Wawra, Phys. Rev. B **61**, 5571 (2000).
52. Y. Zhuang, V. Holý, J. Stangl, A. A. Darhuber, P. Mikulik, S. Zerlauth, F. Schäffler, G. Bauer, N. Darowski, D. Lübbert, et al., J. Phys. D **32**, A224 (1999).
53. G. T. Baumbach, D. Lübbert, and M. Gailhanou, J. Phys. D **32**, A208 (1999).
54. Q. Shen and S. Kycia, Phys. Rev. B **55**, 15791 (1997).
55. Q. Shen, S. W. Kycia, E. Tentarelli, W. Schaff, and L. F. Eastman, Phys. Rev. B **54**, 16381 (1996).
56. M. Hanke, H. Raidt, R. Köhler, and H. Wawra, Appl. Phys. Lett., submitted (2003).
57. M. Meixner (2001), Ph.D. thesis, Technische Universität Berlin (2001).
58. E. Schöll and S. Bose, Solid-State Electron. **42**, 1587 (1998).
59. M. Baus and J. Colot, Phys. Rev. A **36**, 3912 (1987).

5 Dynamical Scattering Effects at Grazing Incidence Conditions

In the previous chapter, X-ray diffuse scattering from free-standing nanoscale islands was simulated exclusively by using the simple kinematical theory. The corresponding experiments were carried out at large angles of incidence α_{i} compared with the critical angle of total external reflection so that refraction effects and the influence of the specularly reflected wave could be neglected. However, for grazing-incidence/exit scattering geometries such as GISAXS and GID these dynamical effects have to be taken into account, and the simulations have to be performed in the framework of the distorted-wave Born approximation. In order to visualize these dynamical effects, the model system of LPE-grown SiGe islands described in Chap. 4 was again chosen. These islands are especially suitable, owing to their high perfection regarding uniform shape and size. Moreover, they are large enough to give a sufficiently high scattering signal. This enables us to qualitatively and quantitatively check the validity of the DWBA approach.

5.1 Grazing-Incidence Small-Angle X-Ray Scattering

Kinematical simulations of GISAXS such as those shown in Fig. 4.8 do not depend on the angle of incidence α_{i}. However, as discussed in Sect. 2.5, refraction and reflection at the surface are not taken into account. Both effects can be properly incorporated by using the five-channel scattering approach (2.97) introduced in Sect. 2.5.2. Since the Fresnel reflection and transmission coefficients $r_{\mathrm{i,f}}$ and $t_{\mathrm{i,f}}$, respectively, enter into the relative intensities of the various scattering channels, the largest influence is expected at angles of incidence that are close to the critical angle of total external reflection. The impact of these effects on the GISAXS intensity distribution is most pronounced for out-of-plane scattering, i.e. when the scattering plane contains the q_z-axis. Here, refraction and also diffuse scattering originating from the specular beam are relevant (see Fig. 2.14), while they are negligible under in-plane scattering conditions. Therefore, it was not necessary to consider dynamical effects for the discussion of lateral island–island correlation in Sect. 4.7.2.

A set of experimental GISAXS intensity patterns recorded at different angles of incidence is shown in Figs. 5.1 and 5.2. Here, truncated SiGe pyramids with an approximate base width of $w = 130\,\mathrm{nm}$ and height $h = 65\,\mathrm{nm}$ were

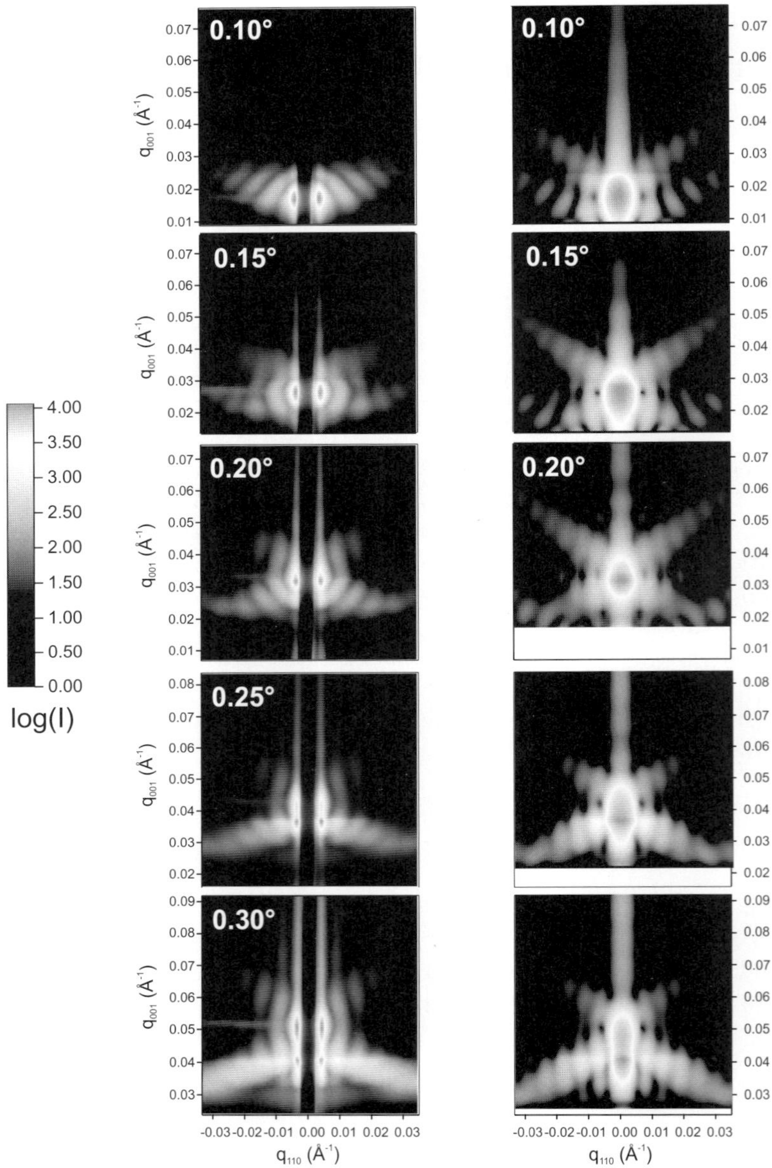

Fig. 5.1. Experimental measurements (*left*) and corresponding theoretical simulations (*right*) of the GISAXS diffuse intensity from SiGe islands (w = 130 nm, h = 65 nm) at different angles of incidence ($\alpha_\mathrm{i} = 0.1°$–$0.3°$). The measurements were performed with the use of a CCD detector at $\lambda = 1.28$ Å (critical angle $\alpha_\mathrm{c} = 0.18°$). The specular beam and the strong 00ℓ CTR at $q_{110} = 0$ were attenuated by a thin metal wire. The simulations were carried out in the framework of the distorted-wave Born approximation (2.97) and do not take into account spatial correlation between the islands [1]

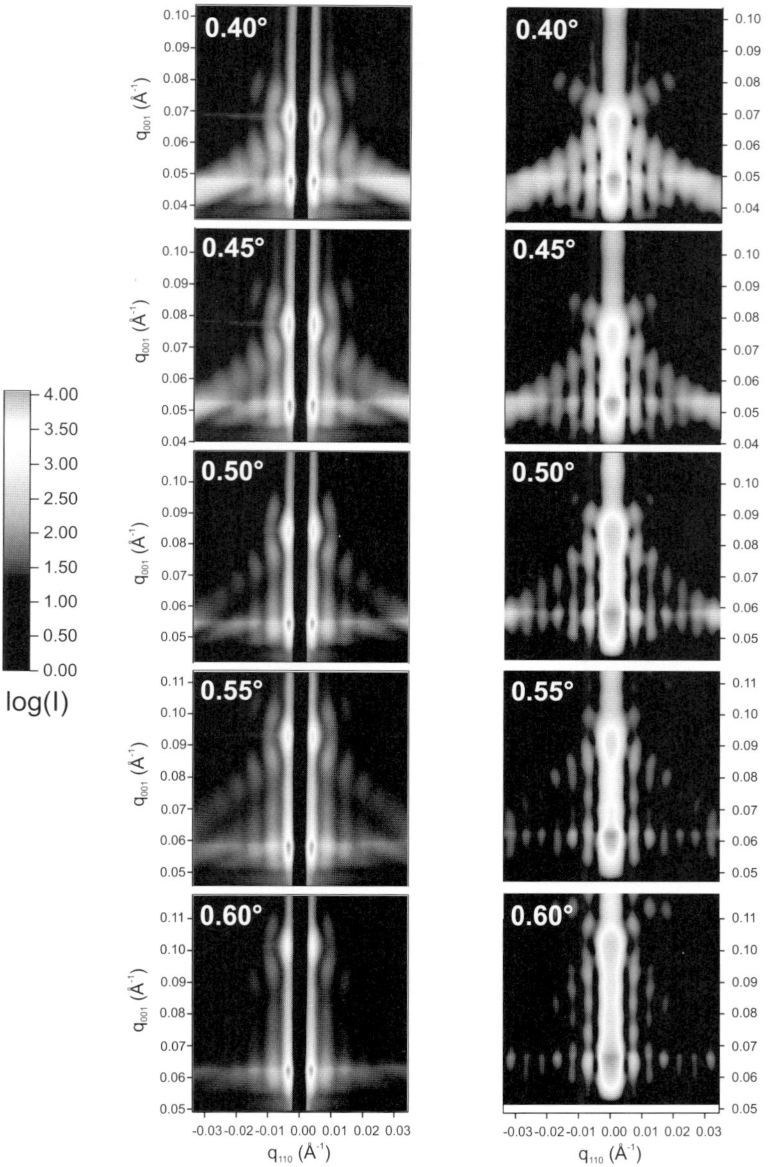

Fig. 5.2. Experimental measurements (*left*) and corresponding theoretical simulations (*right*) of the GISAXS diffuse intensity from SiGe islands (w = 130 nm, h = 65 nm) at different angles of incidence (α_i = 0.4°–0.6°). The measurements were performed with the use of a CCD detector at λ = 1.28 Å (critical angle α_c = 0.18°). The specular beam and the strong 00ℓ CTR at q_{110} = 0 were attenuated by a thin metal wire. The simulations were carried out in the framework of the distorted-wave Born approximation (2.97) and do not take into account spatial correlation between the islands [1]

investigated. The measurements were performed at the high-brilliance ID11 beamline at ESRF using a very well-collimated and narrow incident beam. A CCD detector was placed at a distance of more than 4 m from the sample. These experimental conditions ensured very good resolution and counting statistics.

The intensity patterns are displayed in the q_y–q_z plane, which is identical to the q_{110}–q_{001} plane. However, since a CCD detector was used, the intensity pattern is not exactly defined within the q_y–q_z plane but is, strictly speaking, defined on the Ewald sphere of the scattered wave vector $\boldsymbol{k}_\mathrm{f}$, which is a bowed area in reciprocal space. The simulation was calculated on the Ewald sphere too. But in any case, since all angles are small, the Ewald sphere and the q_y–q_z plane are nearly identical.

A clear dependence of the experimental GISAXS intensity distributions on the angle of incidence α_i can be seen, proving the important role of dynamical scattering. The experimentally observed behavior can be reproduced by calculations using the five-channel DWBA approach as described in Sect. 2.5.2, and there is excellent agreement between experiment and theory. In particular, the diffuse scattering in the vicinity of the specular beam is clearly reproduced in the simulations.

At low angles of incidence (Fig. 5.1), the diffuse scattering in the vicinity of the Yoneda wing (originating from the Fresnel transmission coefficient t_f) and the diffuse intensity around the specular beam overlap and cannot be clearly separated. With increasing angle of incidence (Fig. 5.2), the diffuse intensity that is related to the specular beam moves to higher values of q_z (relative to the Yoneda wing). Both the specular peak and the Yoneda wings show $\{111\}$ truncation rods (CTRs) that are modulated by fringes which are related to the finite size of the islands. Owing to refraction effects, the Yoneda CTR is bowed and crosses the q_z axis at a finite value of q_z (not at $q_z = 0$ as expected from kinematical scattering theory). In contrast, the CTRs from the specular beam appear at an angle of $\pm 54.7°$.

The experimental and theoretical data shown here are in very good correspondence with each other, and only slight deviations can be observed:

1. It is noticeable that the simulated 00ℓ truncation rod is rather broad, while the experimental CTR (which has been attenuated by a metal wire in the experiment) is very sharp. This finite-size effect is due to the restricted volume of the sample that was used in the simulations. The horizontal substrate dimensions were three times the island base width, and the vertical dimension was similar to the island height.
2. Owing to the cell size (about eight times the Si lattice parameter) that was used in the simulations, weak artificial intensity appears at large values of α_i and large values of q_{001} (Fig. 5.2).
3. Note that only the diffuse scattering was calculated by the DWBA. Therefore, a sharp specular beam is missing in the simulations.

4. Sharp satellite rods at about $q_y = 0.003\,\text{Å}^{-1}$ are visible in the experimental data. These rods are created by spatial correlation between neighboring islands and imply a mean spacing along $\langle 110 \rangle$ of about $\langle d_{110} \rangle = 210\,\text{nm}$. Spatial correlation was, however, not considered in the simulations.
5. The experimental intensity pattern is in a sense washed out as compared with the simulations, which are more rich in detail. This could be induced by a finite island size distribution. However, since the island array is rather dense, effects of multiple refraction/scattering could also be responsible. The picture schematically sketched in Fig. 2.14 represents an idealized case. In reality, the scattered and/or reflected beams have to pass through other, additional islands before they can leave the sample. This may lead to absorption and multiple-refraction effects.

The problems addressed in the first two items can, in principle, solved by using smaller cell sizes in the numerical calculations and by increasing the sample volume, i.e. the volume of the substrate. However, each of the calculations shown here requires about 6 hours on a very fast computer.[1] Using half the cell size and double the substrate dimensions would lead to unfeasible computing times in the order of several hundreds of hours.

The last item is still under investigation. Recent experiments with samples with a lower island area coverage have not substantially changed the results shown in Figs. 5.1 and 5.2.

5.2 Grazing-Incidence Diffraction

The theoretical treatment of grazing-incidence diffraction requires numerical simulations using the DWBA approach discussed in Sect. 2.5.3. It has been pointed out that scattering from free-standing islands requires the consideration of two waves (the incident and specularly reflected waves), whereas all four waves that are present inside the crystal are necessary to treat diffuse scattering from the substrate.

Let us first focus on experimental data on the in-plane diffuse intensity in the vicinity of the 220 reciprocal-lattice point which has been recorded close to the critical angle of total external reflection. These measurements were performed at the ID10 beamline at ESRF in a triple-crystal arrangement as sketched in Fig. 3.6a. The experimental intensity pattern displayed in Fig. 5.3a looks very complicated, and it is indeed determined by a complex interplay among the shape, size, and strain field inside and in the proximity of the SiGe island. In the radial direction, three strong peaks can be observed at about $q_{\text{radial}} = 3.259\,\text{Å}^{-1}$, $3.247\,\text{Å}^{-1}$, and $3.239\,\text{Å}^{-1}$. These are related to horizontally dilated areas inside the island. At decreasing values of q_{radial},

[1] A Pentium IV processor with a clock frequency of 1 GHz was employed.

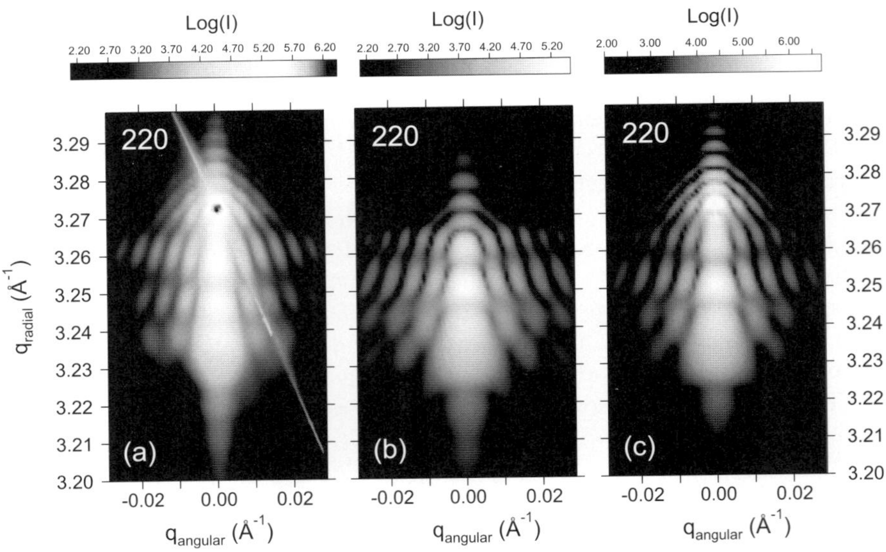

Fig. 5.3. Diffuse intensity in the vicinity of the 220 in-plane reciprocal-lattice point: (**a**) experiment, (**b**) simulation (island only), (**c**) simulation (island and substrate). The angle of incidence $\alpha_\mathrm{i} = 0.25°$ was chosen to be slightly above the critical angle of total external reflection (for Si, $\alpha_\mathrm{c} = 0.22°$ at $\lambda = 1.542$ Å). The simulations were based on the DWBA and used the island model shown in Fig. 4.11b. The simulated intensity was integrated vertically over $\Delta q_z = 0.02$ Å^{-1} [2]

increasing horizontal strain is probed. As can be seen in Fig. 4.10, the horizontal strain component ε_{xx} is a strictly monotonic function of the height z inside the island. Therefore, the diffuse scattering at low values of q_radial can be related to regions close to the top of the island, whereas at high values of q_radial scattering from the base of the island is found. This is also substantiated by the pronounced fringes that are generated by the finite horizontal size of the island. At low values of q_radial, the distance between the fringes is large, thus implying a small horizontal extension in real space, i.e. the top of the island is probed. In contrast, the distance between the fringes decreases on going to larger values of q_radial, thus implying a large horizontal extension in real space. The corresponding areas are located at the island base.

It is interesting to distinguish between the scattering signal from the island itself and the signal originating from diffuse scattering inside the underlying substrate. As can be seen in Fig. 5.3b,c, the diffuse intensities from the island and substrate are almost separated in reciprocal space. At low values of q_radial, the scattering signal arises from strained areas inside the island, whereas the diffuse intensity from the substrate is mainly concentrated at values around $q_\mathrm{radial} = 3.272$ Å^{-1}. Note that the frequencies of the fringe pattern at large values of q_radial are different for the different calculations. The fringes are determined not by the actual horizontal size of the island but by the lateral extension of the strain field inside the substrate. This fact should enable future investigation of the strain field that penetrates from the island into the underlying substrate.

Although the lateral strain tensor component ε_{xx} is a smooth function of z, three distinct *peaks* are seen in the diffuse scattering along q_{radial}. The origin of these peaks is a phase effect and can be explained simply. Let us first restrict ourselves to the direct scattering channel. On the basis of (2.104a), the exponential function has maxima when the Bragg condition is locally fulfilled in-plane, and the local horizontal atomic period can be evaluated accordingly. On the other hand, the exponential function gives additional maxima when $q_\perp = q_z = 2\pi n/z$, where n is an integer value and z the height inside the island.

For reasons of simplicity, a linear increase of the lateral lattice spacing with increasing values of z (Fig. 5.4a) is assumed here. This is certainly a rather coarse approximation, since the lateral strain tensor component ε_{xx} is not a linear function of z. Moreover, ε_{xx} does not depend on z only, but also depends on x and y (see Fig. 4.10). However, the smooth, monotonic

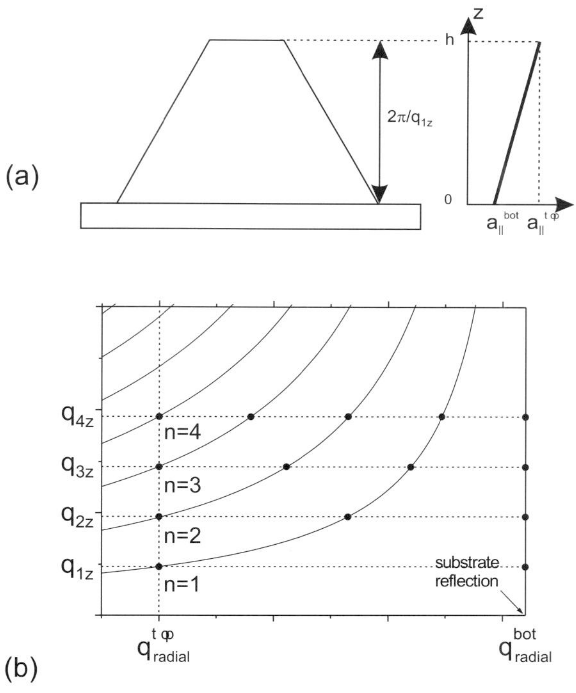

Fig. 5.4. Effect of (**a**) a linear increase of the horizontal lattice parameter $a_\parallel$ with z on (**b**) the positions (q_{radial}, q_z) of the maxima in reciprocal space. The *lines* are given by (5.1). The substrate reflection appears at $q_{\text{radial}}^{\text{bot}} = 2\pi\sqrt{8}/a_\parallel^{\text{bot}}$. The *dots* indicate the maximum intensity at fixed values of q_z

behavior of ε_{xx} as a function of z may justify the linear approximation. For a 220 reflection, the positions (q_z, q_{radial}) of the maxima in reciprocal space are given by

$$q_z = \frac{2\pi n(a_{\parallel}^{\text{top}} - a_{\parallel}^{\text{bot}})}{h} \left(\frac{2\pi\sqrt{8}}{q_{\text{radial}}} - a_{\parallel}^{\text{bot}} \right)^{-1} . \tag{5.1}$$

The quantities $a_{\parallel}^{\text{top}}$ and $a_{\parallel}^{\text{bot}}$ are the horizontal lattice spacings at the top and the bottom of the island, respectively, and h is the island height. Since the island is coherently grown on the substrate, $a_{\parallel}^{\text{bot}}$ can be identified as the Si lattice parameter.

The expression given in (5.1) is displayed in Fig. 5.4b. From this figure, it is clear that the number of maxima along q_{radial} that are visible in X-ray diffuse scattering increases with increasing q_z. The maxima at a given value of q_z are equidistantly spaced[2] with respect to q_{radial}. This is a direct consequence of the linearly increasing horizontal strain inside the island depicted in Fig. 5.4a. Any deviation from this linear law should thus lead to nonequidistantly spaced maxima positions. Indeed, the complex strain tensor field inside a three-dimensional island does not justify the assumption of a linear increase. This can be seen in Fig. 4.10b for an SiGe island with a homogeneous Ge composition. Therefore, the diffuse scattering from a homogeneous island should show a somewhat different behavior for the peak positions from the simple formula given in (5.1). However, additional deviations can be also induced by a Ge chemical composition gradient inside the island.

Indeed, the maxima positions observed in the experiment are not equidistantly spaced. Therefore, FEM calculations were undertaken [2] which employed different island models, as shown in Figs. 4.11 and 4.12. The result for the strain field was then used to simulate the X-ray diffuse scattering in the framework of the DWBA as described in Sect. 2.5.3. Such calculations are shown in Fig. 5.3. Again the island model with an abrupt change of the Ge concentration at about one-third of the island height has been used.

In Figs. 5.5 and 5.6, experimental and simulated X-ray diffuse scattering maps in the q_{radial}–q_z plane are shown. Note that only the scattering from the island itself is depicted, and diffuse scattering from the substrate has been omitted. The data set shows a pronounced influence of α_{i} on the diffuse intensity pattern. Let us first discuss large angles of incidence, e.g. $\alpha_{\text{i}} = 0.60°$. In both the experimental data and the simulations, the shape of the diffuse intensity is similar to the iso-intensity lines sketched in Fig. 5.4, proving the qualitative correctness of the simple model used here. At large angles α_{i} the corresponding q_z values are large, and according to Fig. 5.4 narrow peaks along q_{radial} appear. These are visible in the simulation, but cannot

[2] From (5.1), the spacings of the maxima are constant for $\varepsilon_{\parallel}^{\max} := (a_{\parallel}^{\text{top}} - a_{\parallel}^{\text{bot}})/a_{\parallel}^{\text{bot}} \ll q_z(2\pi/h)$.

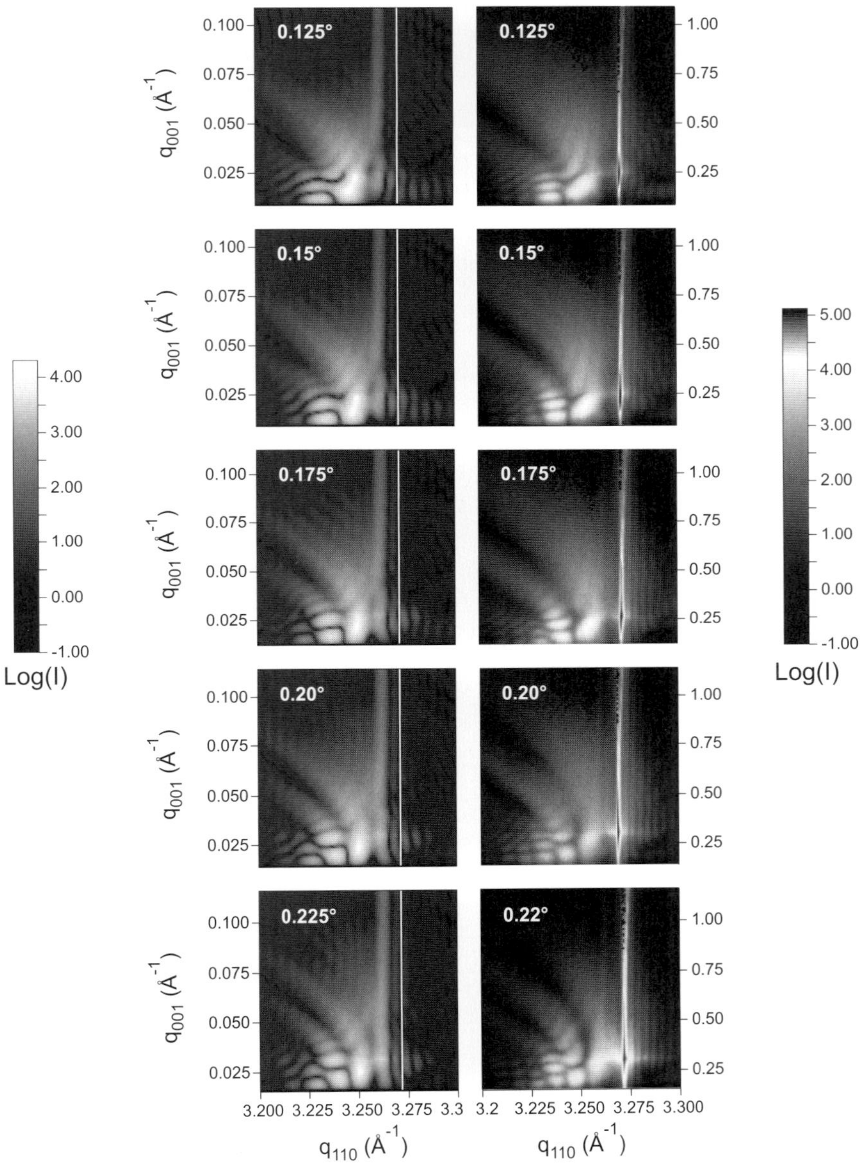

Fig. 5.5. Experimental measurements (*right*) and corresponding theoretical simulations (*left*) of the diffuse intensity (GID, 220) of SiGe islands at different angles of incidence ($\alpha_i = 0.125°$–$0.225°$). The scattering plane contains the $\langle 110 \rangle$ direction. The simulations were carried out in the framework of the distorted-wave Born approximation. They contain the scattering signal from the island only and used the island model shown in Fig. 4.11b. The positions of the substrate peak and CTR are marked by *vertical white lines* [2]

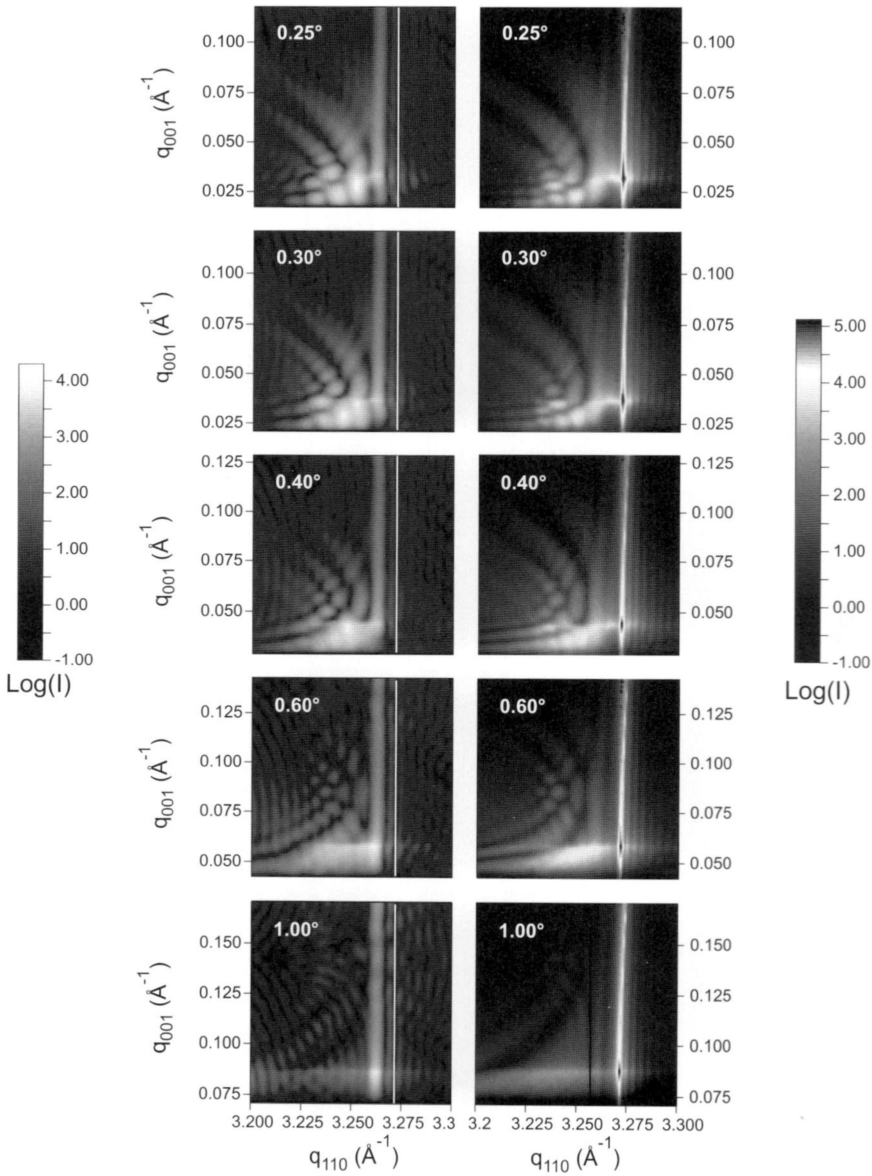

Fig. 5.6. Experimental measurements (*right*) and corresponding theoretical simulations (*left*) of the diffuse intensity (GID, 220) of SiGe islands at different angles of incidence ($\alpha_i = 0.25°–1.00°$). The scattering plane contains the $\langle 110 \rangle$ direction. The simulations were carried out in the framework of the distorted-wave Born approximation. They contain the scattering signal from the island only and used the island model shown in Fig. 4.11b. The positions of the substrate peak and CTR are marked by *vertical white lines* [2]

be resolved further in the experiment. For decreasing values of α_i, the peaks along q_{radial} shift outwards until only two peaks are present at $\alpha_i = 0.125°$.

At low values of α_i the crystal truncation rods of the $\{111\}$ facets of the islands become strong, leading to a characteristic tail that is inclined by 54.7° with respect to the q_z axis. Moreover, at smaller angles of incidence, not only direct scattering from the island but also all other scattering channels become important. This leads to additional structure in the diffuse scattering. The entire pattern of X-ray diffuse scattering now becomes very complicated and cannot be interpreted by a simple picture. However, there is very good agreement between the experiment and the theoretical simulations in Figs. 5.5 and 5.6, which proves the excellent quality of the DWBA for describing the actual scattering scenario for free-standing islands.

In the work described in this section, the finite-element method was used to calculate the strain field inside mesoscopic structures. It is worth mentioning that a different approach, called *isostrain scattering*, has been developed by Kegel et al. [3, 4, 5]. This method is more direct, in the sense that the chemical composition can be directly deduced from the scattered intensities. However, this approach works for cylindrically symmetric islands only and can thus not be applied to truncated pyramids with a square base. Though the authors make use of a four-channel scattering process, scattering from the substrate is omitted.

References

1. M. Schmidbauer, D. Grigoriev, M. Hanke, P. Schäfer, T. Wiebach, and R. Köhler, in preparation (2003).
2. D. Grigoriev, M. Hanke, M. Schmidbauer, P. Schäfer, O. Konovalov, and R. Köhler, J. Phys. D **36**, A225 (2003).
3. I. Kegel, T. H. Metzger, A. Lorke, J. Peisl, J. Stangl, G. Bauer, K. Nordlund, W. V. Schoenfeld, and P. M. Petroff, Phys. Rev. B **63**, 035318 (2001).
4. I. Kegel, T. H. Metzger, A. Lorke, J. Peisl, J. Stangl, G. Bauer, J. M. Garcia, and P. M. Petroff, Phys. Rev. Lett. **85**, 1694 (2000).
5. I. Kegel, T. H. Metzger, P. Fratzl, J. Peisl, A. Lorke, J. M. Garcia, and P. M. Petroff, Europhys. Lett. **45**, 222 (1999).

6 Characterization of Quantum Dots

This chapter deals with the investigation of the diffuse scattering from semiconductor quantum dot ensembles. We focus on two III–V compound materials, namely InP/InGaP and InGaAs/GaAs. On the basis of these systems, we try demonstrate the capability of X-ray diffuse scattering to obtain information about the shape, size, lateral and vertical positional correlation, and chemical composition of the quantum dots. It turns out that InP/InGaP quantum dots grown by gas phase molecular-beam epitaxy show a distinct shape asymmetry. It will be demonstrated that X-ray diffuse scattering along with corresponding simulations can determine the three-dimensional aspect ratios of the base widths to the height. These quantities are often used as input for the calculation of exciton binding energies. InGaAs/GaAs quantum dots grown by metal-organic chemical vapor deposition do not exhibit such a shape asymmetry. Here, X-ray scattering experiments along with corresponding simulations have been used to determine the shape and density of the buried quantum dots.

6.1 Structural Properties of Quantum Dots

Structural properties such as size, shape, elastic strain, and its spatial distribution strongly influence the optical properties of semiconductor quantum dots [1]. In principle, the electronic energy levels and corresponding optical matrix elements are determined by the specific shape of the dot (e.g. an asymmetric shape), and its associated anisotropic strain field inside the quantum dot and the surrounding barrier layers. For reasons of feasibility, however, a fixed quantum dot shape is often used in theoretical calculations of strain and electronic structure, and the dependence of the optical properties on simple geometric parameters is discussed. Among these parameters are the vertical aspect ratio $Q = w/2h$ of the island base width w and the height h (see [2, 3]), and the lateral aspect ratio w_1/w_2, reflecting different quantum dot base widths w_1 and w_2 in different lateral directions [4] (Fig. 6.1).

The characterization, understanding, and control of the size and shape of quantum dots are important not only for a fundamental understanding of the optical properties but also for future applications in optical de-

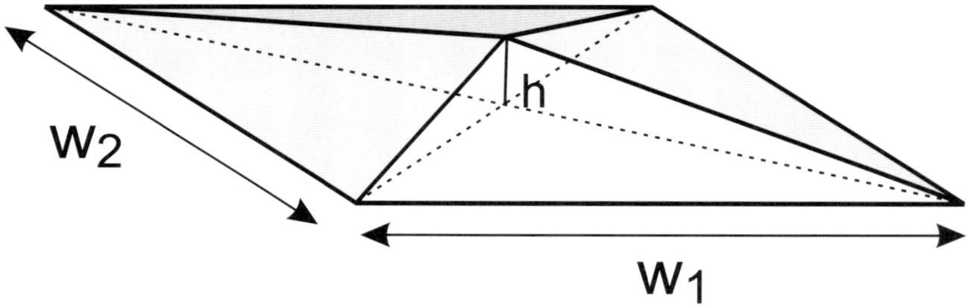

Fig. 6.1. Geometric parameters w_1, w_2, and h of a quantum dot, here assumed to be a pyramid

vices. In III–V material systems such as InAs/GaAs or InP/InGaP, the most popular way of generating quantum dots is through the use of the Stranski–Krastanow growth mode [5]. It is generally accepted that the equilibrium shape of such self-assembled strained islands is determined by the balance of surface free energy and elastic strain energy. Since real surfaces are often strongly anisotropic, significant deviations from highly symmetrical islands (e.g. hemispherical quantum dots or lens-shaped rotationally symmetric quantum dots [6]) may appear. This anisotropy in surface free energy may cause faceting of the island surface, resulting in a complicated quantum dot shape [7]. The situation can be further complicated by kinetic limitations during growth, e.g. an anisotropic surface diffusion length [8] may also have a large influence on the quantum dot shape. Consequently, a large variety of quantum dot shapes has been reported experimentally [4, 9, 10, 11, 12, 13, 14]. These observations were accompanied by various theoretical studies [15, 16, 17, 18, 19] calculating the equilibrium shape of a quantum dot and its dependence on growth kinetics.

On the basis of the atomic structure of III–V materials, the surface energies related to the [110] and $[1\bar{1}0]$ directions are expected to differ, with a consequent effect on the shape of the quantum dots, while the [100] and [010] directions are equivalent. In this chapter, results obtained for two different III–V material systems will be presented. In Sects. 6.2 and 6.3, we concentrate on InP quantum dots that are grown on (001) $In_{0.48}Ga_{0.52}P$/GaAs, which exhibit a pronounced shape asymmetry with respect to the [110] and $[1\bar{1}0]$ directions [20]. Surprisingly, no shape asymmetry can be observed in $In_xGa_{1-x}As$ quantum dots grown by metal-organic chemical vapor deposition (MOCVD) on (001) GaAs (Sect. 6.4). In this ternary system, the indium content x or, more accurately, the indium composition profile is a very critical parameter since it affects decisively the electronic and optical properties. In many experimental studies, a variety of different indium profiles has been observed [21, 22, 23].

6.2 Free-Standing InP Quantum Dots

6.2.1 Surface Morphology

Free-standing InP quantum dots grown by gas phase molecular-beam epitaxy [24] on $In_{0.48}Ga_{0.52}P$/GaAs show a distinct shape asymmetry. An atomic force micrograph, displayed in Fig. 6.2, clearly shows the elongated shape of the InP quantum dots along [110]. A statistical analysis of this and other AFM images yields average island base widths along [110] and $[1\bar{1}0]$ of about $w_1 := w_{110} = 45\,\text{nm} \pm 10\,\text{nm}$ and $w_2 := w_{1\bar{1}0} = 30\,\text{nm} \pm 10\,\text{nm}$, respectively, with an average height of about $h = 8\,\text{nm}$. Cross-sectional transmission electron microscopy (TEM) reveals flat, lens-shaped objects or flat, faceted objects rather than steep, domelike features (Fig. 6.3a, [24]). Also, scanning electron microscopy (SEM) of free-standing quantum dots gives some hints of a weak shape anisotropy, although the contrast is poor. Surprisingly, on-zone bright-field images in plan-view TEM seem to show InP quantum dots with their edges aligned parallel to the $\langle 100 \rangle$ directions (Fig. 6.3b, [24]). However, as demonstrated by Liao et al. [25], the interpretation of these images are not straightforward but requires multiple-beam dynamical image simulations.

In order to avoid misinterpretation, it is necessary to determine the crystallographic orientation of the GaAs substrate. The [110] and $[1\bar{1}0]$ directions can be distinguished by evaluating the orientation of oval defects [26] of the GaAs substrate.

The InP quantum dots reported here are elongated along the [110] direction. In contrast, Jin-Phillipp and Phillipp [27] have observed InP quantum dots that extend along the $[1\bar{1}0]$ direction with a higher lateral aspect ratio

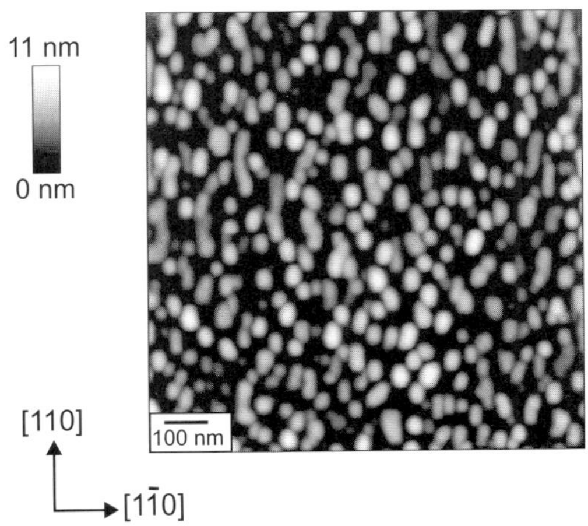

Fig. 6.2. Atomic force micrograph of free-standing InP quantum dots grown on (001) $In_{0.48}Ga_{0.52}P$/GaAs [20]

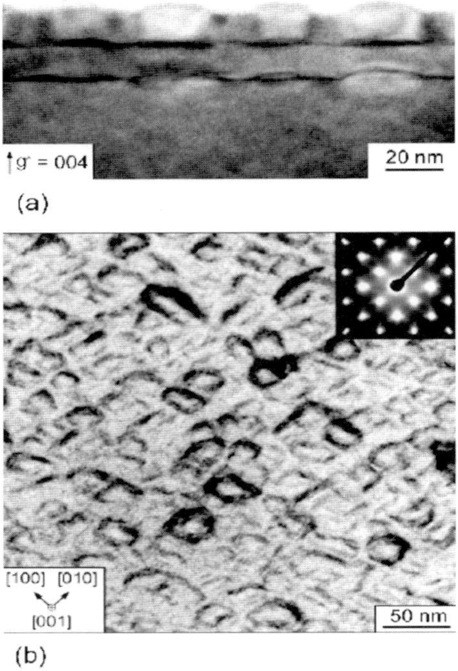

Fig. 6.3. Transmission electron microscope images of InP/$In_{0.48}Ga_{0.52}P$ quantum dots grown on (001)GaAs: (**a**) cross-sectional view, (**b**) plan view [24]

of $w_2/w_1 = 60\,\mathrm{nm} : 20\,\mathrm{nm}$. Also, Sugisaki et al. [28] found InP islands which were elongated along $[1\bar{1}0]$, but with a rather small lateral aspect ratio.

The reasons for the elongation of the InP quantum dots along [110] have not yet been clarified and need theoretical modeling. The equilibrium island shape has been determined by density functional theory (DFT). These calculations [17] show islands elongated along [110] with a base width geometrical aspect ratio of about 3:2, thus fitting the experimental observations.

Another important parameter that can modify growth is the surface miscut, which has been shown to be important in determining the shape anisotropy. At sufficiently large values of typically 2°, the size, shape, and positional correlation of quantum dots can be manipulated through the terrace width of the substrate (see e.g. [29]). The terrace width, on the other hand, depends on the state of step bunching on the sample surface, which can be changed by employing different growth conditions and different buffer layer thicknesses. The GaAs substrates that were used in the work described in this book do not exhibit any significant surface miscut. It is smaller than 0.05°, as determined by X-ray diffraction. This small value of 0.05° corresponds to a terrace width of at least 300 nm, which is far too high to influence the 3D growth of the InP islands. Therefore, the small miscut of $\approx 0.05°$ is not expected to play a significant role. A possible reason for the elongated shape could be kinetic limitations during growth. It is well known from kinetic Monte Carlo simulations [8] that an anisotropic diffusion length leads to elongated islands. It is possible that different diffusion lengths can be induced by the (2×1) surface reconstruction.

In order to obtain a statistically significant measure of the size asymmetry, grazing-incidence small-angle X-ray scattering was performed. The elliptically shaped intensity distribution shown in Fig. 6.4 reflects the asymmetric quantum dot shape. To a good approximation, the horizontal aspect ratio of the ellipse is related to the island base widths along [110] and $[1\bar{1}0]$ via $\Delta q_{1\bar{1}0}/\Delta q_{110} = w_{110}/w_{1\bar{1}0} \approx 1.5$, which is in good agreement with the geometrical aspect ratio estimated by AFM. Superimposed on the ellipse in Fig. 6.4 are four strong peaks that are related to positional correlation along the lateral $\langle 100 \rangle$ directions. The evolution of positional correlation as a function of InP deposition is described in more detail in [24] and will not be discussed further here.

6.2.2 Experimental Results

We shall now focus on the question of how the shape influences the strain relaxation inside a quantum dot. From elasticity theory, it is known [30] that the strain distribution inside a quantum dot is, for a given fixed shape, independent of the quantum dot size. However, as will be addressed here,

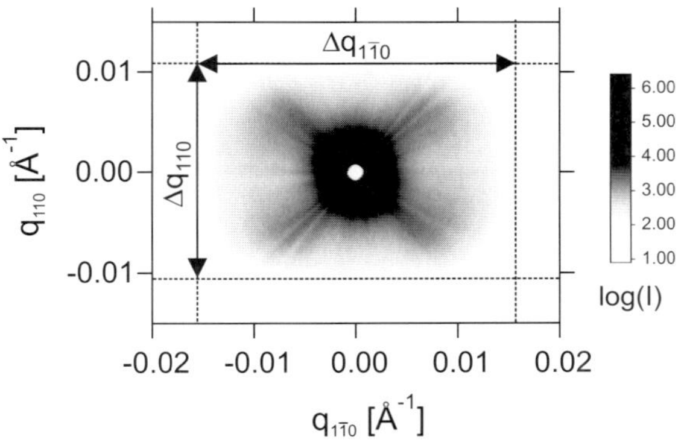

Fig. 6.4. In-plane grazing-incidence small-angle X-ray scattering from the sample shown in Fig. 6.2. The elliptically shaped intensity distribution with its main axis along $[1\bar{1}0]$ is a consequence of the elongated quantum dot shape along [110]. The aspect ratio of the ellipse $\Delta q_{1\bar{1}0}/\Delta q_{110}$ in reciprocal space scales inversely with the corresponding lateral aspect ratio $w_{1\bar{1}0}/w_{110}$ in real space. Four peaks are superimposed on the ellipse, which are due to positional correlation along the $\langle 100 \rangle$ directions [20]

it is critically sensitive to the shape. The strain distribution may, therefore, serve as a sensitive fingerprint of the quantum dot shape.

Again, the indirect method was used where the experimental X-ray scattering pattern was compared with corresponding kinematical simulations using the finite-element method. Since structural properties such as shape, size, and chemical composition enter sensitively into these simulations, the calculations can be refined until satisfactory agreement with experiment is achieved. A detailed description of this method is given in Sect. A.2.

The rather moderate lattice mismatch compared with InAs/GaAs, of 3.8% between InP ($a = 5.87$ Å) and $In_{0.48}Ga_{0.52}P$ ($a = 5.65$ Å) ensures that linear elasticity theory can be used to a very good approximation. The continuum elastic properties of III–V materials exhibit fourfold symmetry, and, in particular, the [110] and [$1\bar{1}0$] directions are treated equivalently. Nevertheless, different strain relaxations in these two directions may occur owing to an asymmetric quantum dot shape. This shape-induced anisotropic strain relaxation is much stronger than the effects that are caused by taking account of the exact atomistic two-fold symmetry as discussed by Pryor et al. [31].

In order to obtain complete information, a variety of reflections was investigated with consistent results. In Fig. 6.5a and 6.5b, the experimental X-ray diffuse intensity in the vicinity of the $1\bar{1}3$ and 113 reciprocal-lattice points, respectively, is displayed for free-standing InP quantum dots. The diffuse intensity is widely spread out in reciprocal space, and the peak position is shifted by $\delta q_{1\bar{1}0}$ and δq_{110} with respect to the corresponding values $q_{1\bar{1}0}$ and q_{110} of the $In_{0.48}Ga_{0.52}P$ buffer reflections (marked by vertical white lines in Fig. 6.5). These shifts are, to a good approximation, related to the corresponding mean values of the total elastic strain components through

$$\langle \varepsilon_{110} \rangle = -\frac{\delta q_{110}}{q_{110}} , \qquad (6.1a)$$

$$\langle \varepsilon_{1\bar{1}0} \rangle = -\frac{\delta q_{1\bar{1}0}}{q_{1\bar{1}0}} , \qquad (6.1b)$$

where ε_{110} and $\varepsilon_{1\bar{1}0}$ are the diagonal elements of the strain tensor ε_{ij} with respect to the [110] and [$1\bar{1}0$] directions, respectively. Note that the total strain is used here, i.e. the $In_{0.48}Ga_{0.52}P$ buffer (which is lattice-matched to GaAs) is used as the reference everywhere.

Evaluation of Fig. 6.5a,b yields values of $\langle \varepsilon_{1\bar{1}0} \rangle = (9.2 \pm 0.5) \times 10^{-3}$ and $\langle \varepsilon_{110} \rangle = (5.0 \pm 0.5) \times 10^{-3}$. It is striking that the elastic relaxation is different for the [110] and [$1\bar{1}0$] directions. This is related to the elongated shape of the quantum dots along the [110] direction. Since the influence of the underlying GaAs substrate and $In_{0.48}Ga_{0.52}P$ buffer is larger in the elongated direction than along the shorter direction, the lattice can relax more efficiently along [$1\bar{1}0$] than along [110]. On the other hand, as depicted in Fig. 6.6a,b, the diffuse intensities in the vicinity of the 404 and 044 reciprocal-lattice points

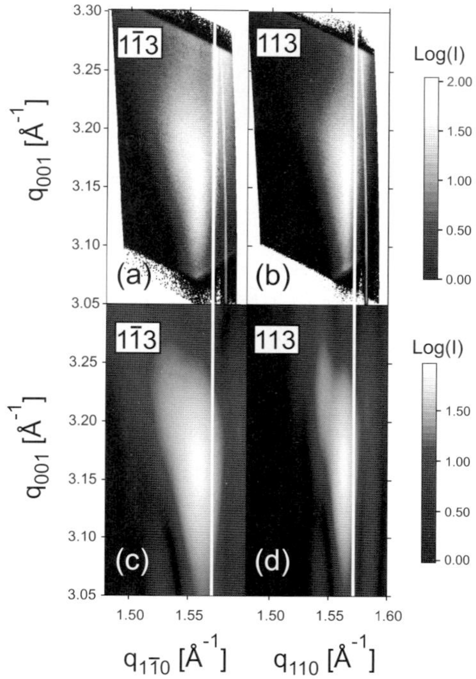

Fig. 6.5. X-ray diffuse scattering of free-standing InP quantum dots in the vicinity of the $1\bar{1}3$ and 113 reciprocal-lattice points: (**a**,**b**) experiment; (**c**,**d**) corresponding simulations. The intensity scale varies logarithmically from *black* to *white* as indicated at the *right*. The Bragg peaks of the $In_{0.48}Ga_{0.52}P$ buffer layer and GaAs substrate are not shown here, since they do not reveal information about the InP quantum dots. They are located at $q_{001} = 3.335\,\text{Å}^{-1}$ and $q_{110} = q_{1\bar{1}0} = 1.572\,\text{Å}^{-1}$, as marked by *vertical white lines*. These correspond to the crystal truncation rods of the sample surface [20]

are identical, proving that the strain field is identical along the [100] and [010] directions.

6.2.3 Simulation of X-Ray Diffuse Scattering

A detailed understanding of the experimental data requires careful theoretical modeling. First, a certain island model, with a certain shape, size, and lattice mismatch, is created. Secondly, the elastic strain field inside and around the island is calculated numerically by using the finite-element method. Finally, this strain field is used to calculate the diffuse X-ray scattering within the kinematical scattering approach. In sequential steps, the shape and size are refined until satisfactory agreement with experiment is achieved.

It is perhaps interesting to note that the simulations can be fitted to the experimental data only by using a somewhat reduced lattice mismatch of

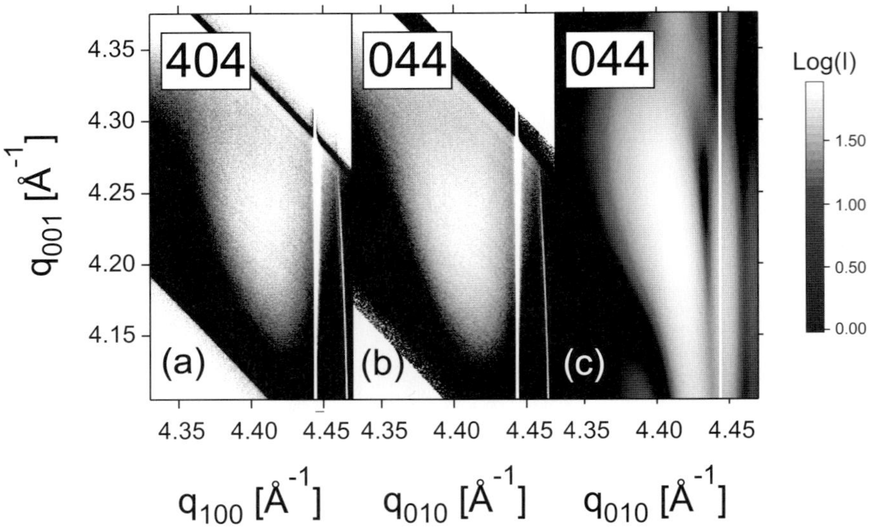

Fig. 6.6. Measured X-ray diffuse scattering of free-standing InP quantum dots in the vicinity of the (**a**) 404 and (**b**) 044 reciprocal-lattice points; (**c**) simulation. The close similarity of (a) and (b) proves that there is no strain anisotropy with respect to the [100] and [010] directions. The intensity scale varies logarithmically from *black* to *white*. The Bragg peaks of the $In_{0.48}Ga_{0.52}P$ buffer layer and GaAs substrate (not shown here) are located at $q_{001} = 4.445\,\text{Å}^{-1}$ and $q_{100} = q_{010} = 4.445\,\text{Å}^{-1}$, as marked by *vertical white lines*. These correspond to the crystal truncation rods of the sample surface [20]

$\Delta a/a = 2.58\%$ between the quantum dots and the underlying $In_{0.48}Ga_{0.52}P$ buffer. It is thus very likely that intermixing of In and Ga takes place during growth. The reduced mismatch corresponds to $In_{0.83}Ga_{0.17}P$ quantum dots. The influence of a possible chemical composition gradient of In inside the quantum dots was not considered in this study, and the quantum dots used in the simulations were modeled as having a homogeneous chemical composition.

For islands with steep side faces (e.g. truncated pyramids with steep facets), the simulations and experiment are at clear variance. In contrast, the use of quantum dots with shallow faces leads to much better results. Therefore, the simulations discussed further here used flat-faceted pyramids with a rectangular base. The remaining free parameters of the simulations were the quantum dot base widths w_1 and w_2 along [110] and $[1\bar{1}0]$, and the quantum dot height h. The simulations turned out to be very sensitive to these parameters, and very good agreement with experiment was achieved for $w_1 = w_{110} = 50\,\text{nm} \pm 5\,\text{nm}$, $w_2 = w_{1\bar{1}0} = 30\,\text{nm} \pm 5\,\text{nm}$, and $h = 7\,\text{nm} \pm 1\,\text{nm}$, which is in good agreement with AFM.

The corresponding calculated X-ray diffuse scattering around the $1\bar{1}3$, 113, and 044 reciprocal-lattice points is displayed in Figs. 6.5c, 6.5d, and 6.6c, respectively. As can be seen in these figures, the simulations reproduce the overall shape of the experimental diffuse intensity, proving that the calculated elastic strain field matches the real one. In particular, the absolute positions of the maxima of the diffuse intensity distribution in reciprocal space are, within the error bars given, identical in the experiment and the simulations. Therefore, the experimentally determined and simulated values of the mean strength of the asymmetry $\langle\alpha\rangle$, as given by

$$\langle\alpha\rangle = \frac{\langle\varepsilon_{1\bar{1}0}\rangle}{\langle\varepsilon_{110}\rangle}, \tag{6.2}$$

agree within the estimated uncertainties: experimentally, $\langle\alpha\rangle = 1.84 \pm 0.10$ is found, and the simulated value is $\langle\alpha\rangle = 1.71 \pm 0.10$.

6.2.4 Strain Relaxation Versus Geometrical Aspect Ratios

Since the quantum dot size (at fixed shape) does not influence the strain distribution, only two of the parameters w_1, w_2, and h are independent. The two remaining relevant quantities which determine the strain distribution are the geometrical aspect ratios, defined as

$$Q_1 = \frac{w_1}{2h}, \tag{6.3a}$$

$$Q_2 = \frac{w_2}{2h}. \tag{6.3b}$$

When the quantum dot height h is increased at fixed base widths w_1 and w_2, the strain inside the quantum dot may relax more efficiently. This is demonstrated in Fig. 6.7a. Here the mean total strains $\langle\varepsilon_{1\bar{1}0}\rangle$ and $\langle\varepsilon_{110}\rangle$, which were evaluated from the peak positions in simulated X-ray diffuse-scattering maps, are plotted as a function of the island height h. In the results of our simulations, the mean lateral strain tensor components $\langle\varepsilon_{1\bar{1}0}\rangle$ and $\langle\varepsilon_{110}\rangle$ increase linearly as a function of the island height. On the other hand, at fixed quantum dot height h, the mean total strain components $\langle\varepsilon_{110}\rangle$ and $\langle\varepsilon_{1\bar{1}0}\rangle$ scale inversely with the quantum dot base widths w_1 and w_2, respectively. As a consequence, $\langle\alpha\rangle$ varies linearly with the lateral aspect ratio Q_1/Q_2 (Fig. 6.7). This useful scaling behavior facilitates a systematic refinement of the simulation.

Although they appear quite reasonable, the linear scaling laws for the *mean* total strains (i.e. averaged over the entire island) plotted in Fig. 6.7 are not obvious. This assertion can be demonstrated by finite-element calculations. We first focus on a pyramid with $w_1 = 50$ nm, $w_2 = 30$ nm, and $h = 7$ nm. In Fig. 6.8, calculated values of ε_{110} and $\varepsilon_{1\bar{1}0}$ are shown in the $(1\bar{1}0)$ and (110) planes, respectively, through the symmetry axis of the pyramid. The calculations show that ε_{110} and $\varepsilon_{1\bar{1}0}$ are different inside the pyramid and the underlying $In_{0.48}Ga_{0.52}P$ buffer, and inside the pyramid they increase

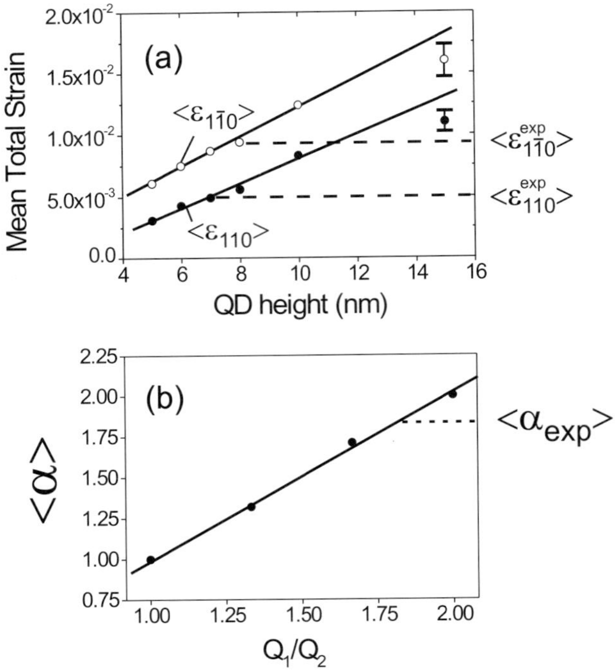

Fig. 6.7. (**a**) Mean total strain, evaluated from calculated X-ray diffuse-scattering maps similar to Fig. 6.5c,d, inside free-standing InP quantum dots (ε_{110}, $\bullet$; $\varepsilon_{1\bar{1}0}$, $\circ$) as a function of the quantum dot height h. Quantum dot base widths of $w_1 =$ 50 nm and w_2 = 30 nm have been used. (**b**) Calculated mean asymmetry ratio $\langle\alpha\rangle = \langle\varepsilon_{1\bar{1}0}\rangle/\langle\varepsilon_{110}\rangle$ ($\bullet$) plotted versus the lateral aspect ratio Q_1/Q_2 of the quantum dot. The *horizontal dashed lines* represent corresponding experimental values taken from Figs. 6.5a,b. The *solid lines* are guides for the eye [20]

monotonically for larger values of z. This means that areas inside the island close to $z = 0$ are strongly stressed and are not able to relax horizontally. Only with increasing position z above the wetting layer may elastic relaxation take place to some extent. This behavior is also illustrated in selected line scans for $\varepsilon_{110}(z)$, $\varepsilon_{1\bar{1}0}(z)$, and $\alpha(z) = \varepsilon_{1\bar{1}0}(z)/\varepsilon_{110}(z)$ along the symmetry axis of the pyramid (Fig. 6.9a). It is interesting that the island is totally relaxed nowhere, i.e. ε_{110} and $\varepsilon_{1\bar{1}0}$ are always smaller than the lattice mismatch $\Delta a/a = 2.58\%$ between the pyramid and the buffer. As a consequence the ratio $\alpha(z)$ is nearly constant for $z \geq 0$. This behavior, which can be observed for island heights h up to about 10 nm, is the main reason for the linear scaling law plotted in Fig. 6.7 for the *mean* total strains. At sufficiently large island height h, however, the situation is different. This is illustrated in Fig. 6.9b, where corresponding line scans for a pyramid with $w_1 = 50$ nm, w_2 = 30 nm, and $h = 15$ nm are shown. Here, the curves for $\varepsilon_{110}(z)$ and $\varepsilon_{1\bar{1}0}(z)$

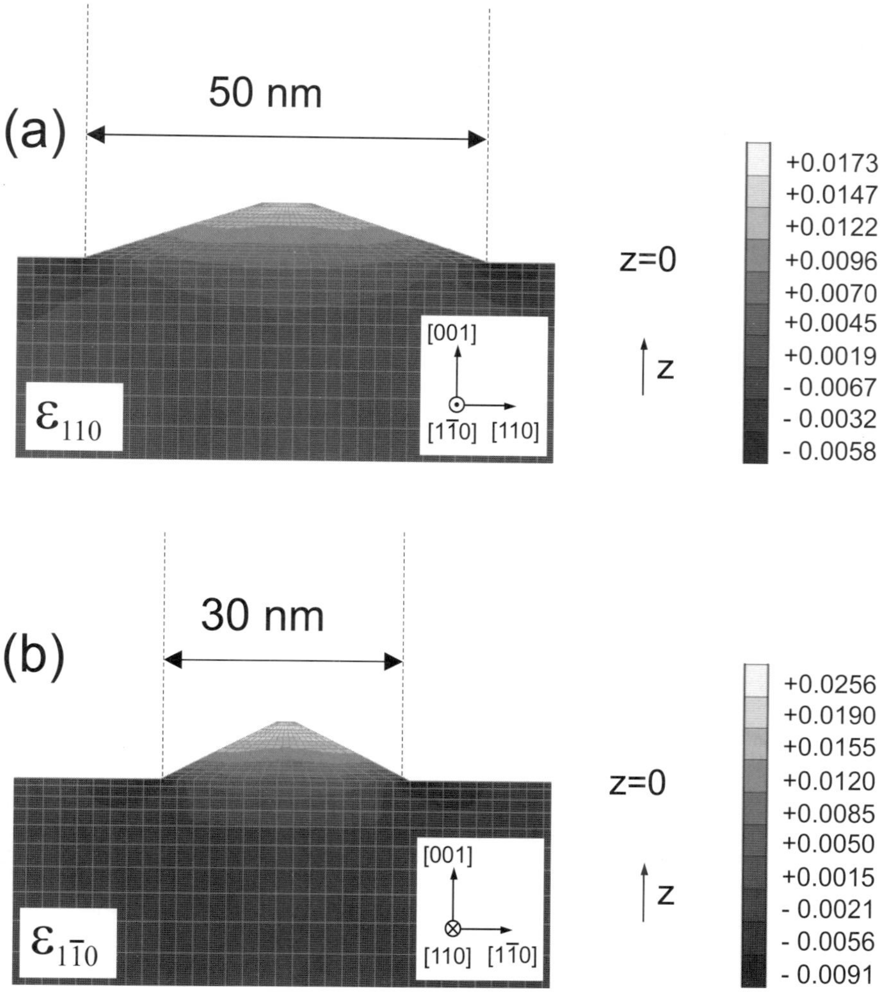

Fig. 6.8. Calculated strain tensor components (**a**) ε_{110} and (**b**) $\varepsilon_{1\bar{1}0}$ in the $(1\bar{1}0)$ and (110) planes, respectively, through the symmetry axis of a strained InP pyramid ($w_1 = 50$ nm, $w_2 = 30$ nm, $h = 7$ nm, lattice mismatch 2.58%) attached to an $In_{0.48}Ga_{0.52}P$ buffer layer.The finite-element method was used [20]

merge at the top of the pyramid, and, consequently, $\alpha(z)$ varies locally inside the island. This is also reflected in Fig. 6.7a for the mean total strains, which show significant deviations from the linear scaling at $h = 15$ nm.

6.2.5 Size and Shape Homogeneity

The calculated diffuse scattering exhibits pronounced shape-induced diffuse scattering that gives rise to intensity oscillations (fringes). One can expect,

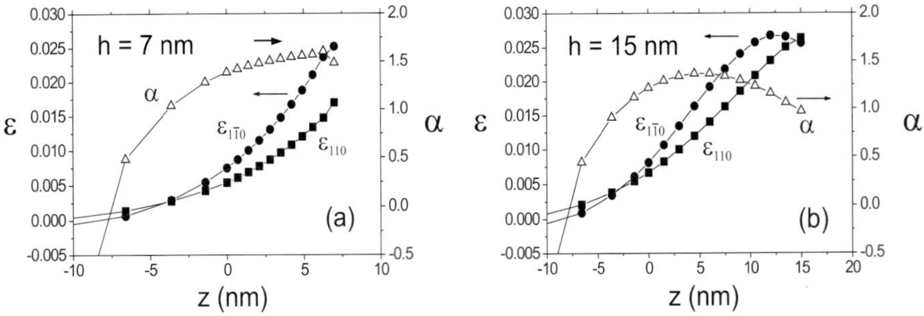

Fig. 6.9. Calculated strain tensor components: (**a**) selected line scans through the symmetry axis of the pyramid shown in Fig. 6.8: $\varepsilon_{110}(z)$ (*filled squares*), $\varepsilon_{1\bar{1}0}(z)$ (*filled circles*), and $\alpha = \varepsilon_{1\bar{1}0}/\varepsilon_{110}$ (*open triangles*). (**b**) Corresponding line scans through a pyramid with $w_1 = 50\,\text{nm}$, $w_2 = 30\,\text{nm}$, $h = 15\,\text{nm}$ [20]

however, that these fringes may not be observable in the experimental data, owing to size fluctuations of the quantum dots and, in fact, no fringes are experimentally observed. As pointed out before, the strain field scales solely with the geometric aspect ratios Q_1 and Q_2. Therefore, a possible quantum dot size distribution does not change the distribution of strain-induced X-ray diffuse scattering in reciprocal space as long as the shape is not changed. The good agreement between experiment and simulation is, therefore, indicative of a rather uniform shape of the quantum dots, although the size fluctuations (15% rms) estimated by AFM are significant.

For a faceted pyramid with rectangular base, the evaluated values of $Q_1 = 3.57$ and $Q_2 = 2.14$ correspond to quantum dot side faces which are tilted by 15.6° and 25.0° with respect to the (001) surface and are thus close to {115} (15.8°) and {113} (25.2°) facets, respectively.

6.3 Buried InP Quantum Dots

6.3.1 Polarization Anisotropy of Photoluminescence

The optical matrix elements and thus the photoluminescence (PL) spectra of quantum dots often show a pronounced polarization anisotropy. In order to study the effect of the asymmetric dot shape and the anisotropic strain relaxation, the lateral linear polarization of the PL from capped InP quantum dots was investigated. Since the dots are elongated along the [110] direction, polarization anisotropy between the [110] and [1$\bar{1}$0] directions is expected [4, 32, 33]. Indeed, the photoluminescence (Fig. 6.10) reveals a significant polarization anisotropy of

$$p = \frac{I_{[110]} - I_{[1\bar{1}0]}}{I_{[110]} + I_{[1\bar{1}0]}} \approx 36\% \,, \tag{6.4}$$

where $I_{[110]}$ and $I_{[1\bar{1}0]}$ denote the integrated PL intensities polarized parallel to the [110] and [1$\bar{1}$0] directions, respectively. No PL polarization anisotropy was observed with respect to the [100] and [010] directions, and also the polarization anisotropy of the $In_{0.48}Ga_{0.52}P$ buffer and GaAs substrate vanished.

Interpretation of the PL data requires detailed theoretical investigations. It is known from such investigations that the optical polarization anisotropy is caused by mixing of the heavy-hole and light-hole states [34]. This coupling can be induced by asymmetric lateral confinement and anisotropic lateral strain [33]. The asymmetric lateral confinement due to an asymmetric quantum dot shape yields an increased lateral polarization in the direction of elongation of the quantum dot [4], i.e. in this case the [110] direction. On the other hand, as discussed above in detail, lateral strain relaxation is more efficient along the shorter axis of the quantum dot, leading to a reduced lateral polarization along the [1$\bar{1}$0] direction. Consequently, both the differences in lateral confinement and the strain anisotropy enhance the degree of polarization along [110].

Anisotropic lateral strain inside the quantum dots is, however, not necessarily connected to an asymmetric quantum dot shape. This has been recently shown by Sugisaki et al. [28], who observed a strong polarization dependence of the PL for InP quantum dots that have nearly identical island base widths along [110] and [1$\bar{1}$0]. Moreover, strong anisotropy was also found for the PL signal from the $In_{0.48}Ga_{0.52}P$ matrix. The authors of [35] claim that the polarization anisotropy is induced by anisotropic strain in the $In_{0.48}Ga_{0.52}P$

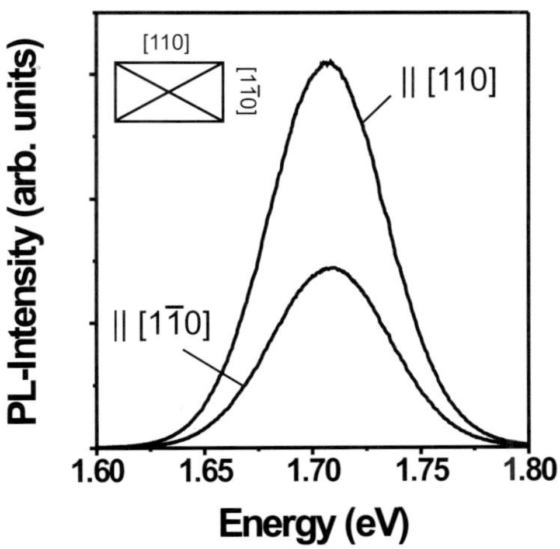

Fig. 6.10. PL spectra (10 K) of buried InP quantum dots polarized linearly along [110] and [1$\bar{1}$0]. The orientation, and the different island base widths along [110] and [1$\bar{1}$0] have been sketched schematically [20]

barrier layer, which is due to the formation of horizontal composition variations along the [110] and [1$\bar{1}$0] directions. Since a polarization dependence was not observed for the $In_{0.48}Ga_{0.52}P$ buffer, this interpretation can be excluded for the samples investigated in the present work. Instead, the measured polarization anisotropy is induced by the elongation of the dots along [110]. This result, therefore, also proves that the quantum dots are still asymmetric with respect to [110] and [1$\bar{1}$0] even after they have been overgrown with $In_{0.48}$ $Ga_{0.52}P$.

6.3.2 Shape and Strain Relaxation

The fact that the shape asymmetry is preserved after overgrowth is also visible in the X-ray diffuse scattering. This is demonstrated in Fig. 6.11 for a single layer of buried InP quantum dots. The strain relaxation is again more pronounced in the [1$\bar{1}$0] direction (Fig. 6.11, left) than along [110] (Fig. 6.11, right), with a mean strength of the asymmetry of about $\langle\alpha\rangle = 1.7 \pm 0.10$, which fits well to the experimental value of $\langle\alpha\rangle \approx 1.84 \pm 0.10$ that is found for the free-standing islands. Moreover, it is striking that the intensity distribution for the buried quantum dots (Fig. 6.11) is quite similar to that for the free-standing quantum dots (Fig. 6.5): in both cases the diffuse inten-

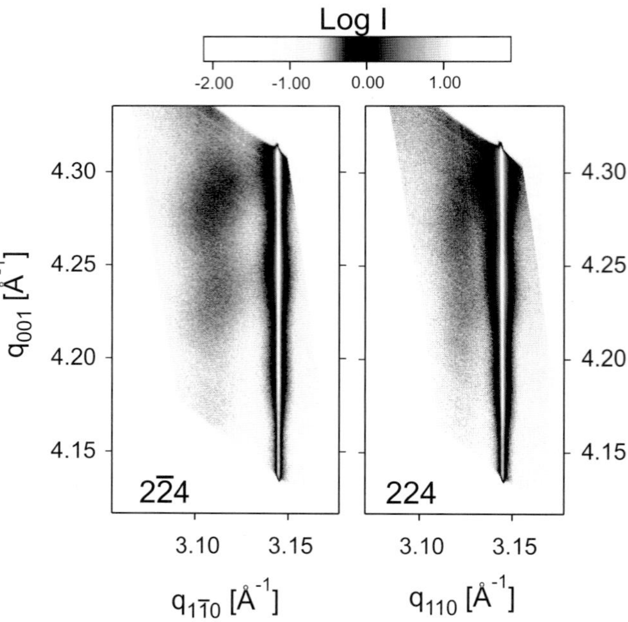

Fig. 6.11. Diffuse scattering from a single layer of buried InP quantum dots in the vicinity of the 224 and 2$\bar{2}$4 reciprocal-lattice points of GaAs. Note that the logarithmic grayscale runs from white to black and back to white

sity appears at the left-hand side of the CTR, i.e. at lower values of q. This implies that mainly dilated parts (with respect to the perfect GaAs and the InGaP buffer layer, i.e. $\varepsilon_{110}, \varepsilon_{1\bar{1}0} > 0$) of the layer are visible in the diffuse scattering. FEM simulations show that, for flat, pyramidally shaped quantum dots, these regions are basically located (i) inside the quantum dot itself, and (ii) in the $In_{0.48}Ga_{0.52}P$ matrix above and below the InP quantum dot. In contrast, compressed areas of $In_{0.48}Ga_{0.52}P$ are located close to the edges of the base of the quantum dot.

The same qualitative behavior can be observed for a vertical stack of InP quantum dots. This is demonstrated in Fig. 6.12 for a 10 period superlattice. Here, the diffuse intensity is concentrated in *resonant diffuse scattering* (RDS) sheets, which are a consequence of vertical correlation of the quantum dot positions inside the quantum dot stack. The RDS sheets are symmetrical with respect to the strain-insensitive angular direction (Fig. 6.12c) but are shifted towards lower values of q_{radial} (Fig. 6.12a,b). Thus, areas inside the multilayer that exhibit horizontal *dilatation* are probed. As can be seen in Fig. 6.12a,b, the lattice dilatation is less pronounced along [110] than along [1$\bar{1}$0]. However, a closer inspection of the diffuse intensity distribution of the RDS sheet in Fig. 6.12b also shows intensity at $q_{\text{rad}} > 3.145\,\text{Å}^{-1}$, indicating compressed areas.

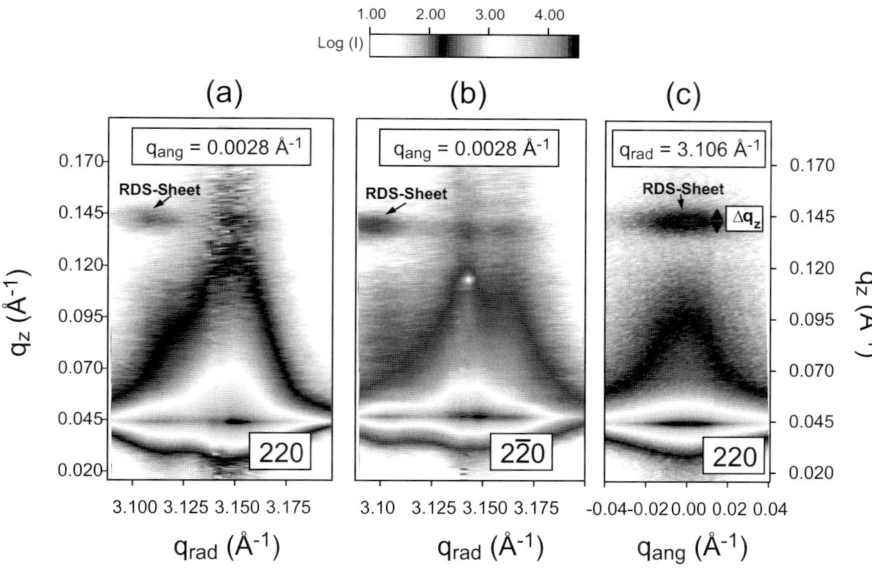

Fig. 6.12. Diffuse scattering of a 10 period InP/$In_{0.48}Ga_{0.52}P$ quantum dot superlattice grown on GaAs(001) in grazing-incidence diffraction geometry: (**a**) 220 radial, (**b**) 2$\bar{2}$0 radial, and (**c**) 220 angular direction. The resonant diffuse scattering (RDS) sheet at $q_z = 0.145\,\text{Å}^{-1}$ corresponds to the first superlattice satellite. Note that the logarithmic grayscale runs from white to black twice [36]

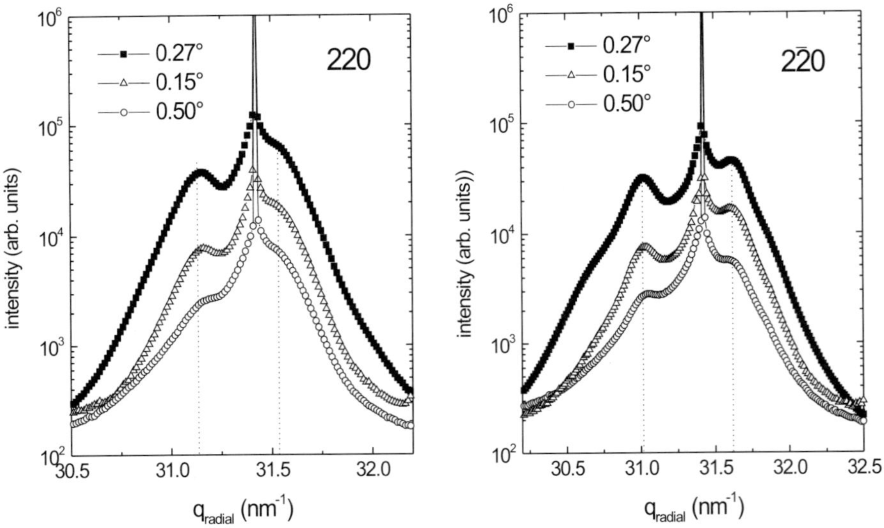

Fig. 6.13. Effect of different penetration depths in triple-crystal grazing-incidence diffraction from a 10 period InP/$In_{0.48}Ga_{0.52}P$ quantum dot superlattice grown on (001)GaAs. Although the intensity of the diffuse scattering changes when the angle of incidence is varied from 0.15° to 0.50°, the relative peak positions with respect to the strong, sharp substrate peak remain unchanged

The mean strength of the asymmetry can be evaluated as $\langle\alpha\rangle = 1.47 \pm 0.05$, which is somewhat reduced compared with the values obtained for free-standing quantum dots and a single layer of buried InP quantum dots. The reasons for this reduced ratio are still not clear. On one hand, the InP thicknesses are not exactly identical. The free-standing and multilayered quantum dots have a nominal thickness of 1.5 nm (5 monolayers InP), whereas the nominal thickness is 1.8 nm (6 monolayers InP) for the single buried layer. These differences are small and are not expected to have a significant influence. On the other hand, another reason for the reduced anisotropy of strain relaxation could be a change of the shape and size of the InP quantum dots during growth, induced by vertical correlation of the lateral quantum dot positions. Tersoff et al. [37] have predicted that vertical stacking of many layers of quantum dots separated by thin spacing layers will enhance the lateral and vertical positional correlation and the size homogeneity. For SiGe/Si multilayers, Mateeva et al. [38] found that merging of islands of different initial size was the dominant mechanism leading to a uniform size distribution. However, Heinrichsdorff et al. [39] observed that the dot size increases in higher layers. This effect can be checked with depth-resolved grazing-incidence diffraction. A variation of the angle of incidence – and thus the penetration depth – does not change the relative peak position of the diffuse scattering with respect to the strong, sharp GaAs substrate peak (see Fig. 6.13). Therefore, it is very unlikely that there is a substantial change of the quantum dot shape and size in higher layers.

6.3.3 Lateral and Vertical Correlation

It is interesting that the GISAXS signal arises mainly from structures at the sample *surface*. Therefore, the signal from free-standing islands at the surface is much stronger than that from similar buried islands. This can be seen in Fig. 6.14, where the GISAXS signal of a 10 period superlattice (see also Fig. 3.7) is compared with GISAXS from a single layer of free-standing InP quantum dots. The low signal can be explained by the small electron density difference between the InP quantum dots and the embedding $In_{0.48}Ga_{0.52}P$ host lattice, whereas a high electron density difference is present for free-standing quantum dots.

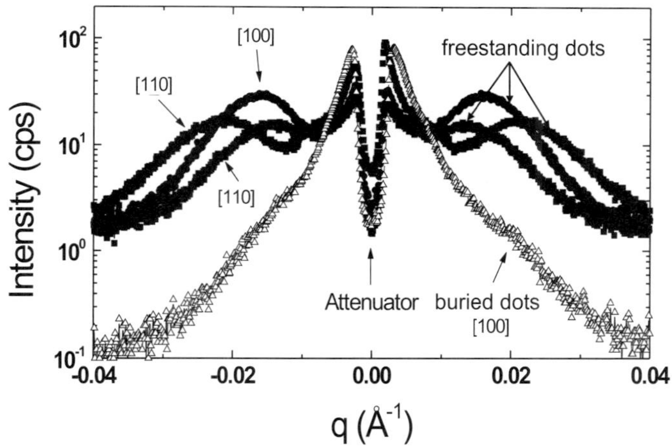

Fig. 6.14. GISAXS of free-standing (*solid symbols*) and buried (*open triangles*) InP quantum dots. The angle of incidence was chosen to $\alpha_i = 0.40°$. The diffuse intensity with the scattering vector $\boldsymbol{q}$ collinear with the [110], [1$\bar{1}$0], and [100] directions is shown. The strong specular beam at $q = 0$ has been attenuated [36]

The lateral and vertical correlation of buried quantum dots can be probed by grazing-incidence diffraction which is – in contrast to GISAXS – additionally sensitive to strains inside the sample. If the quantum dot positions are spatially correlated, not only the electron density but also the strain field undergoes a corresponding spatial modulation. Therefore, the pure electron density contrast visible in GISAXS is enlarged. This is illustrated in Fig. 6.15, where the in-plane intensity distribution for a 10 period $InP/In_{0.48}Ga_{0.52}P$ quantum dot superlattice is shown around the 220 and $2\bar{2}0$ reciprocal-lattice points. The strong central peak at $q_{radial} = 3.144\,Å^{-1}$ arises partly from

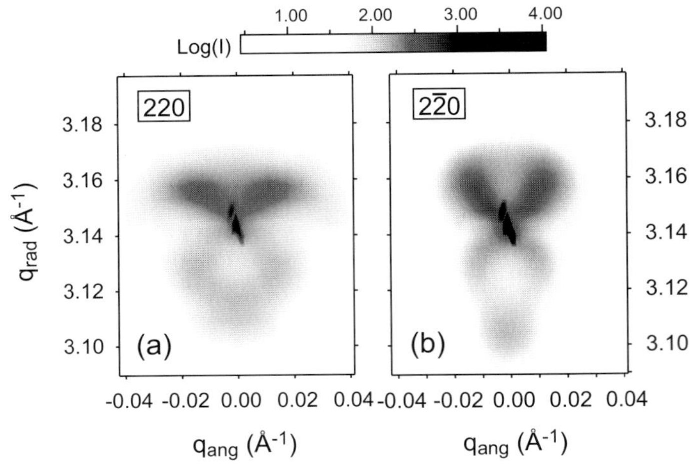

Fig. 6.15. In-plane GID map of a 10 period InP/$In_{0.48}Ga_{0.52}P$ quantum dot superlattice grown on GaAs(001): (**a**) around 220, and (**b**) around $2\bar{2}0$ [36]

scattering from the perfect part of the $In_{0.48}Ga_{0.52}P$ matrix and partly from the underlying GaAs substrate, while the surrounding diffuse scattering is mainly due to strain fluctuations. The diffuse scattering exhibits four correlation peaks that can be attributed to spatial correlation along the $\langle 100 \rangle$ directions.

The in-plane measurements were performed at $\alpha_i = \alpha_f = \alpha_c$. Therefore, mainly the strain field of the $In_{0.48}Ga_{0.52}P$ cap layer near the surface was probed. Obviously, the strain caused by the InP quantum dots penetrates through the upper $In_{0.48}Ga_{0.52}P$ layer to the next interface, where it may create nucleation centers for the following layer of InP quantum dots. Consequently, vertical spatial correlation of the InP quantum dots layers may be induced. Vertical sections are depicted in Fig. 6.12. As can be clearly seen, the diffuse scattering is partly concentrated in a horizontal RDS sheet around the first superlattice satellite ($q_z = 0.145\,\text{Å}^{-1}$), proving strong vertical correlation of the quantum dot layers. Evaluation of the vertical width (FWHM) Δq_z by using (2.56) yields a vertical correlation length of about $\xi = 60\,\text{nm}$, which is commensurate with the total multilayer thickness of 65 nm.

6.4 Buried InGaAs Quantum Dots

6.4.1 Shape Analysis

The actual quantum dot shape depends critically on the growth conditions, and, similarly to InP/InGaP, asymmetric shapes have been found for InAs/GaAs and InGaAs/GaAs quantum dots also. However, the direction and strength of the asymmetry are different. Many papers report a weak shape asymmetry [12, 32, 33] only; however, elongated quantum dots with large aspect ratios both along the [110] direction [4] and along the [1$\bar{1}$0] direction [40, 41] have also been observed. However, a shape asymmetry need not necessarily be present, as will be demonstrated in this section for the example of InGaAs quantum dots grown by MOCVD on GaAs(001) substrates. Here, we focus on five-period $In_xGa_{1-x}As$/GaAs superlattices with typical thicknesses of about $d_{\mathrm{InGaAs}} = 1\,\mathrm{nm}$ and $d_{\mathrm{GaAs}} = 18\,\mathrm{nm}$ and $x = 60\%$.

Figure 6.16 displays transmission electron micrographs of such a five-period InGaAs/GaAs quantum dot superlattice in plan view (Fig. 6.16a) and cross-sectional view (Fig. 6.16b). From these images the quantum dot size can be estimated as $w = 20\,\mathrm{nm}$ and $h = 2$–$5\,\mathrm{nm}$. As can be clearly seen, the quantum dots exhibit a high degree of vertical ordering (even at rather large spacer layer thicknesses of more than 20 nm, where the vertical correlation is believed to vanish [43]). The high degree of vertical correlation can be visualized by X-ray diffraction (Fig. 6.17a,b), which exhibits strong RDS sheets in the vicinity of the in-plane 200 and 020 reciprocal-lattice points. The vertical width of the RDS sheets is identical to the vertical width of the superlattice satellites along the [00ℓ] crystal truncation rod, indicating

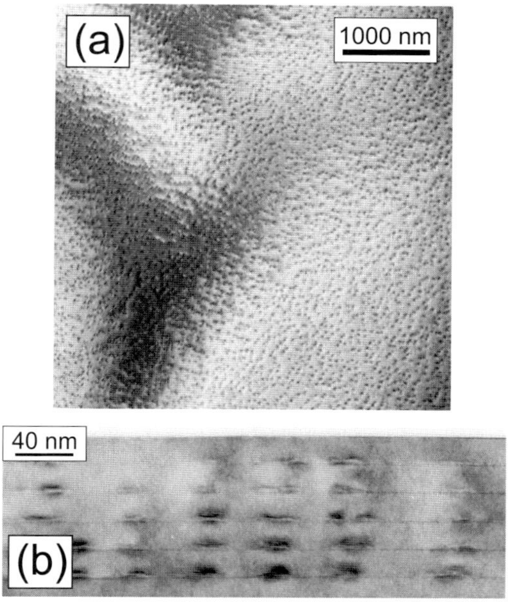

Fig. 6.16. Transmission electron micrographs of an InGaAs/GaAs quantum dot superlattice in plan view (**a**) and cross-sectional view (**b**). The latter shows a high degree of vertical correlation of the quantum dot positions. In contrast, the plan-view image suggests randomly distributed positions of the quantum dots within each individual layer. See also [42]

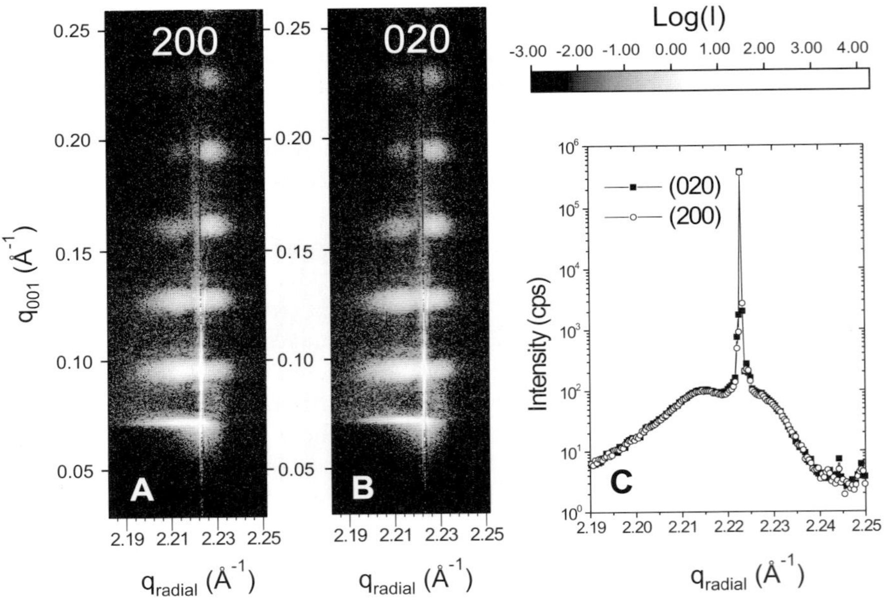

Fig. 6.17. Diffusely scattered intensity (GID) from an InGaAs/GaAs quantum dot superlattice (see Fig. 6.16) around the (**a**) 200 and (**b**) 020 reciprocal-lattice points. The observation of nearly no difference between the two ⟨100⟩ directions (as well for the two ⟨110⟩ directions, not shown here) suggests a symmetric quantum dot shape with respect to these directions. The symmetry can also be seen for the integrated intensities (**c**)

a vertical correlation length which is identical to the entire thickness of the quantum dot superlattice. From the vertical periodicity in reciprocal space of $\Delta q_z = 3.24 \times 10^{-2}\,\text{Å}^{-1}$, the mean superlattice period can be extracted as $t = 19.4\,\text{Å}$.

As can be checked in Fig. 6.17, the InGaAs quantum dots show no size asymmetry with respect to the [100] and [010] directions. Also, the [110] and $[1\bar{1}0]$ directions are equivalent (not shown here).

In order to reveal more detailed information about the lateral and vertical chemical composition profiles, the experimental X-ray data have to be compared with scattering simulations based on the DWBA according to the iterative numerical simulation method described in Sect. 2.3.3. The simulations and experiment were performed for the 220 reflection (Fig. 6.18), since the theoretical approach described in Sect. 2.3.2 does not work properly for quasi-forbidden 200 reflections.

Various quantum dot shapes and chemical composition profiles were used in the simulations, and a 1 nm wetting layer with the same chemical composition as the quantum dots was introduced in all cases. Both dilated and compressed areas are visible in the 220 diffraction pattern. Dilated areas lead

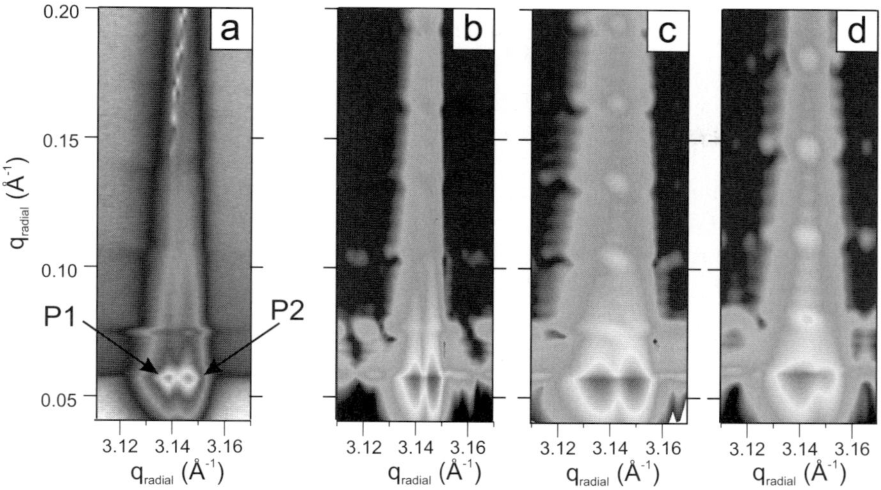

Fig. 6.18. Diffusely scattered intensity (GID) from an InGaAs/GaAs quantum dot superlattice (see Fig. 6.16) around the 220 reciprocal-lattice point. (**a**) Experiment; (**b**)–(**d**) depict various dynamical scattering simulations based on different quantum dot shapes and lateral quantum dot spacings. For details see text. The intensity scale varies from black via white to black again [42]

to enhanced intensity, denoted by P1, at the left side of the CTR, whereas compressed areas lead to enhanced intensity, denoted by P2, at the right side of the CTR. The dilated and compressed areas can probably be attributed to the quantum dot itself and the surrounding spacer layer, respectively. The strength and location of these areas depend strongly on the quantum dot shape. Therefore, although these features P1 and P2 appear qualitatively in all simulations, there are distinct differences regarding their radial position and intensity ratio, especially in Fig. 6.18c,d. Model (c) (Fig. 6.18c) is based on a cuboid with dimensions 20 nm × 20 nm × 5 nm, which approximates the experimentally found flat, truncated pyramidal shape [23], whereas model (d) is based on a pointed pyramid with exactly the same base width, height, and indium composition ($x_{\mathrm{In}} = 0.45\%$). The experimentally observed peak intensity ratio $I_{\mathrm{P1}}/I_{\mathrm{P2}} \approx 1$ is in excellent agreement with a cuboidal quantum dot shape. On the other hand, a pyramidal quantum dot shape results in a pronounced peak P1, which is in clear discordance with experiment. Thus, an important result of this study is that a pointed pyramidal quantum dot shape can be definitely excluded.

While the simulated peak intensity ratio $I_{\mathrm{P1}}/I_{\mathrm{P2}}$ in Fig. 6.18c fits well to the experimental data, it is evident that the simulated peak positions of P1 and P2 are still not correct. These can be adjusted to the experimental data by changing the lateral distance $\langle d \rangle$ between adjacent dots. While $\langle d \rangle$ amounts to 40 nm in Fig. 6.18c, the best agreement is achieved with $\langle d \rangle = 80$ nm, used in Fig. 6.18b. The corresponding quantum dot density of about $2 \times 10^{10}\,\mathrm{cm}^{-2}$ is in good agreement with results from plan-view TEM images.

6.4.2 Chemical Composition

Since the simulations show that the In composition profile does not significantly impact GID-related features, additional investigations in the HRXRD geometry were undertaken. Figure 6.19 shows the experimental intensity distribution in the vicinity of the 224 reciprocal-lattice point. The coherent diffraction from the perfect GaAs substrate is labeled S, and the vertical periodicity of the quantum dot stack leads to vertical intensity modulations (superlattice satellites) along the CTR ($q_{110} = 3.143\,\text{Å}^{-1}$). The intensity clouds located to the left and right of the CTR are due to diffuse scattering. The diffuse intensity is concentrated in RDS sheets, indicating strong vertical correlation of the quantum dots. The vertical distance between the substrate peak S and the zero-order superlattice peak is sensitive to the averaged vertical strain, which is related to the In content. Using a cuboidal quantum dot shape, the In content was gradually modified between $x = 30\%$ and $x = 45\%$ in corresponding simulations (Fig. 6.19b–d). The best agreement

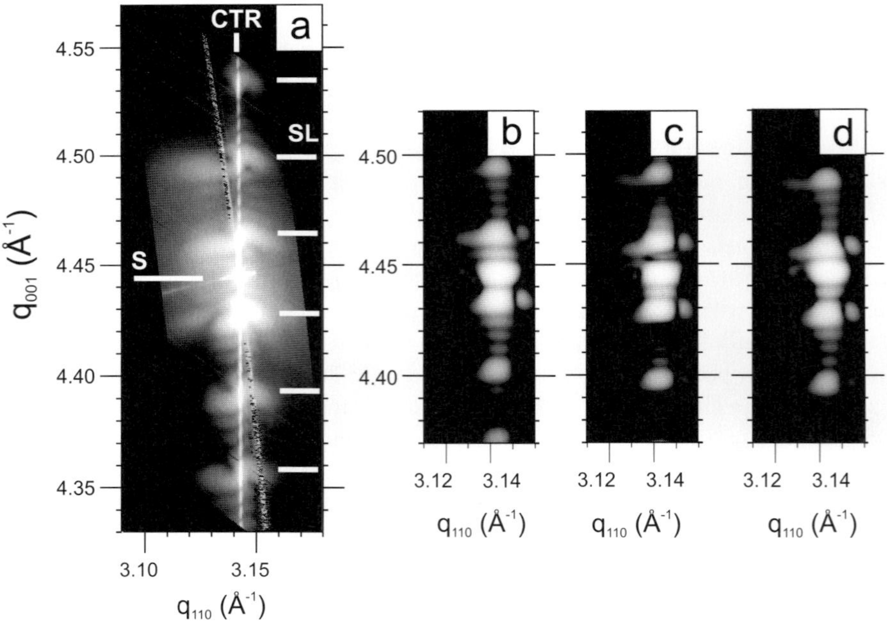

Fig. 6.19. Diffuse intensity from an InGaAs/GaAs quantum dot superlattice (see Fig. 6.16) in the proximity of the asymmetrical 224 reciprocal-lattice point: (**a**) experimental, (**b**)–(**d**) scattering simulations assuming a cuboidal quantum dot shape and In contents of 30%, 40%, and 45% [42], respectively

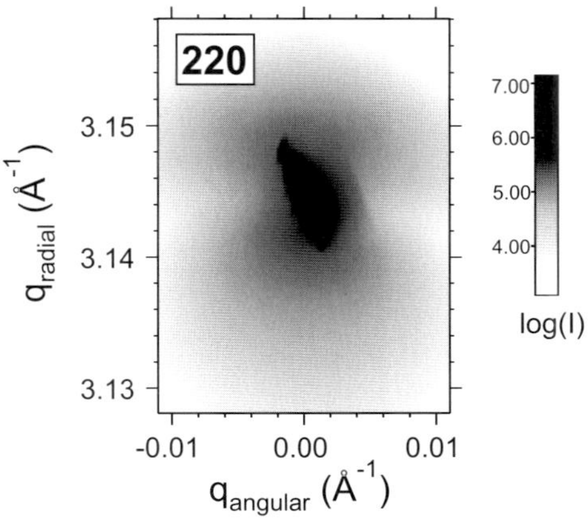

Fig. 6.20. Grazing-incidence diffraction in-plane reciprocal-space map of an InGaAs/GaAs quantum dot superlattice in the vicinity of the 220 reciprocal-lattice point. The rather smooth intensity distribution indicates that the quantum dot positions are horizontally uncorrelated

was achieved assuming an In content of $x = 40\%$, which seemingly represents a remarkable deviation from the nominal value of $x = 60\%$. However, we have to emphasize that the derived In content contains large uncertainties: besides the In composition inside the quantum dots, the average strain inside the entire quantum dot stack is influenced by the the wetting-layer thickness and the quantum dot size, which can be only estimated by TEM. Another decisive parameter is the lateral spacing between the quantum dots, which, however, could be determined rather accurately by X-ray simulations (Fig. 6.18) and plan-view TEM (Fig. 6.16a).

6.4.3 Lateral Correlation

Despite the pronounced vertical correlation, no lateral correlation of the quantum dots could be detected, surprisingly. The 220 in-plane reciprocal-space map shows only a broad intensity distribution of diffuse scattering (Fig. 6.20), without any intensity modulations (correlation peaks) that could be assigned to lateral correlation.

References

1. D. Bimberg, M. Grundmann, and N. N. Ledentsov, *Quantum Dot Heterostructures* (Wiley, Chichester, 1999).
2. M. A. Cusack, P. R. Briddon, and M. Jaros, Phys. Rev. B **56**, 4047 (1997).
3. M. Califano and P. Harrison, J. Appl. Phys. **88**, 5870 (2000).

4. W. Yang, H. Lee, T. Johnson, P. Sercel, and A. Norman, Phys. Rev. B **61**, 2784 (2000).
5. I. Stranski and L. Krastanow, Sitzungsberichte d. Akademie d. Wissenschaften in Wien, Abt. IIb **146**, 797 (1937).
6. I. Kegel, T. H. Metzger, A. Lorke, J. Peisl, J. Stangl, G. Bauer, J. M. Garcia, and P. M. Petroff, Phys. Rev. Lett. **85**, 1694 (2000).
7. C. Walther, R. P. Blum, H. Niehus, W. T. Masselink, and A. Thamm, Phys. Rev. B **60**, R13962 (1999).
8. C. Heyn, Phys. Rev. B **63**, 033403 (2001).
9. C. Walther, W. Hoerstel, H. Niehus, J. Erxmeyer, and W. T. Masselink, J. Cryst. Growth **209**, 572 (2000).
10. P. Ballet, J. B. Smathers, H. Yang, C. L. Workman, and G. J. Salamo, Appl. Phys. Lett. **77**, 3406 (2000).
11. A. Lorke, R. J. Luyken, A. O. Govorov, J. P. Kotthaus, J. M. Garcia, and P. M. Petroff, Phys. Rev. Lett. **84**, 2223 (2000).
12. J. Marquez, L. Geelhaar, and K. Jacobi, Appl. Phys. Lett. **78**, 2309 (2001).
13. S. A. Chaparro, Y. Zhang, and J. Drucker, Appl. Phys. Lett **76**, 3534 (2000).
14. K. Maehashi, N. Yasui, Y. Murase, T. Ota, T. Noma, and H. Nakashima, J. Electron. Mater. **29**, 542 (2000).
15. J. Tersoff and R. M. Tromp, Phys. Rev. Lett. **70**, 2782 (1993).
16. E. Pehlke, N. Moll, A. Kley, and M. Scheffler, Appl. Phys. A **65**, 525 (1997).
17. Q. K. K. Liu, N. Moll, M. Scheffler, and E. Pehlke, Phys. Rev. B **60**, 17008 (1999).
18. L. G. Wang, P. Kratzer, N. Moll, and M. Scheffler, Phys. Rev. B **62**, 1897 (2000).
19. N. Moll, M. Scheffler, and E. Pehlke, Phys. Rev. B **58**, 4566 (1998).
20. M. Schmidbauer, F. Hatami, M. Hanke, P. Schäfer, K. Braune, W. Masselink, R. Köhler, and M. Ramsteiner, Phys. Rev. B **65**, 125320 (2002).
21. N. Liu, H. Lyeo, C. Shih, M. Oshima, T. Mano, and N. Koguchi, Appl. Phys. Lett. **80**, 4345 (2002).
22. N. Liu, J. Tersoff, O. Baklenov, J. Holmes, and C. Shih, Phys. Rev. Lett. **84**, 334 (2000).
23. A. Lenz, R. Timm, H. Eisele, C. Hennig, S. Becker, R. Sellin, U. W. Pohl, D. Bimberg, and M. Dähne, Appl. Phys. Lett. **81**, 5150 (2002).
24. F. Hatami, U. Müller, H. Kissel, K. Braune, R.-P. Blum, S. Rogaschewski, H. Niehus, H. Kirmse, W. Neumann, M. Schmidbauer, et al., J. Cryst. Growth **216**, 26 (2000).
25. X. Z. Liao, J. Zou, X. F. Duan, D. J. H. Cockayne, R. Leon, and C. Lobo, Phys. Rev. B **58**, R4235 (1998).
26. Y. G. Chai and R. Chow, Appl. Phys. Lett. **38**, 796 (1981).
27. N. Y. Jin-Phillipp and F. Phillipp, J. Appl. Phys. **88**, 710 (2000).
28. M. Sugisaki, H. W. Ren, S. V. Nair, K. Nishi, S. Sugou, T. Okuno, and Y. Masumoto, Phys. Rev. B **59**, R5300 (1999).
29. H. J. Kim, Y. J. Park, E. K. Kim, and T. W. Kim, J. Cryst. Growth **223**, 450 (2001).
30. M. Grundmann, O. Stier, and D. Bimberg, Phys. Rev. B **52**, 11969 (1995).
31. C. Pryor, J. Kim, L. W. Wang, A. J. Williamson, and A. Zunger, J. Appl. Phys. **83**, 2548 (1998).

32. Y. Nabetani, T. Ishikawa, S. Noda, and A. Sasaki, J. Appl. Phys. **76**, 347 (1994).
33. N. N. Ledentsov, P. D. Wang, C. M. SotomayorTorres, A. Y. Egorov, M. V. Maximov, V. M. Ustinov, A. E. Zhukov, and P. S. Kop'ev, Phys. Rev. B **50**, 12171 (1994).
34. U. Bockelmann and G. Bastard, Phys. Rev. B **45**, 1688 (1992).
35. Y. Zhang and A. Mascarenhas, Phys. Rev. B **57**, 12245 (1997).
36. M. Schmidbauer, F. Hatami, P. Schäfer, M. Hanke, T. Wiebach, H. Niehus, W. Masselink, and R. Köhler, Mater. Res. Soc. Symp. Proc. **642**, J6.8 (2001).
37. J. Tersoff, C. Teichert, and M. G. Lagally, Phys. Rev. Lett. **76**, 1675 (1996).
38. E. Mateeva, P. Sutter, J. C. Bean, and M. G. Lagally, Appl. Phys. Lett. **71**, 3233 (1997).
39. F. Heinrichsdorff, M. H. Mao, N. Kirstaedter, A. Krost, D. Bimberg, A. O. Kosogov, and P. Werner, Appl. Phys. Lett. **71**, 22 (1997).
40. T. Hanada, B. Koo, H. Totsuka, and T. Yao, Phys. Rev. B **64**, 165307 (2001).
41. W. Ma, R. Nötzel, H.-P. Schönherr, and K. H. Ploog, Appl. Phys. Lett. **79**, 4219 (2001).
42. M. Hanke, D. Grigoriev, M. Schmidbauer, P. Schäfer, R. Köhler, U. W. Pohl, R. Sellin, D. Bimberg, N. D. Zakharov, and P. Werner, to be published in Physica E (2003).
43. Q. Xie, A. Madhukar, P. Chen, and N. Kobayashi, Phys. Rev. Lett. **75**, 2542 (1995).

7 Characterization of Interface Roughness

In this chapter, the characterization of interface roughness is discussed. Roughness can be often described by a self-affine model of the height profile. This holds even in the case of very flat stepped interfaces of semiconductor layer systems grown on vicinal substrates. However, the existence of atomic steps and step bunches induces anisotropic interface morphologies and inclined inheritance of interface roughness in a multilayer. How these effects can be analyzed will be discussed in detail. In particular, a simple geometric model is presented which simplifies the evaluation of inclined interface roughness. We shall also demonstrate that with increasing strain, the model of self-affine roughness will break down, and monodisperse step bunches show up that may partially release their internal strain.

7.1 Roughness in Vicinal Semiconductors

The exciton size and, thus, the optical properties of a quantum well structure are very sensitive to local thickness fluctuations of the wells. These fluctuations are given in a natural way by the interfacial roughness. When the typical lateral dimension of the roughness is in the 10–100 nm range, localized exciton states varying in binding and transition energies may be induced [1, 2]. These give rise to inhomogeneous broadening of the integrated macroscopic photoluminescense (PL) or cathodoluminescence (CL) spectrum. Spatially resolving techniques, e.g. scanning near-field optical microscopy (SNOM)[1] and microphotoluminescence (μPL) [5], allow spatially resolved exciton spectroscopy with a resolution of better than 100 nm. In that case the inhomogeneously broadened PL line of a macroscopic PL spectrum can be resolved into very narrow lines that are directly related to single, localized exciton states. Roughness is therefore a crucial structural parameter that strongly influences the optical properties and requires detailed analysis. Indeed, in the last decade, the surface and interface roughness of thin layers have been extensively investigated [6, 7, 8, 9].

[1] In a SNOM experiment [3, 4], a tapered fiber tip is scanned across a sample, and optical information such as reflectivity, fluorescence, luminescence, or polarization can be derived with subwavelength resolution. In addition, topographical information can usually be obtained in a similar way to atomic force microscopy.

Originally, X-ray diffuse scattering was developed for characterization of the interfacial roughness of amorphous samples [6]. However, a few years later, it was also extensively applied to perfect crystalline multilayers (see e.g. [10, 11, 12, 13, 14]), and some of these studies revealed effects that were not observed for amorphous samples. The main reason for these new effects in the diffuse scattering from interface and surface roughness is the presence of atomic steps, which play a decisive role when the layers are grown on vicinal semiconductor substrates. Usually, the atomic steps are not equidistantly spaced but exhibit a terrace width distribution. In that case the surface (or interface) may be divided into regions with high step densities and regions with few or no steps. This phenomenon is referred to as *step bunching*.

The effect of step bunching has been extensively investigated and can be ascribed to step–step interaction. In particular, Lässig [15] found an interplay between long-range repulsion and short-range attraction. This effect is especially strong for vicinal Si(113) surfaces [16, 17, 18, 19, 20] and vicinal GaAs(113) [21] and GaAs(110) [22] surfaces. In contrast, on vicinal Si(001) and GaAs(001) surfaces, step bunching is not a strong effect. Here, the evolution of roughness is dominated by growth kinetics [23, 24, 25, 26, 27]. These nonequilibrium processes depend strongly on the growth conditions and are still not completely understood. In these cases, mostly irregular step bunches have been observed.

In contrast, elastic strain has a large impact on the step bunching morphology. In the continuous description of this effect [28, 29, 30], the driving force for surface undulation is the reduction of elastic energy at the expense of surface free energy. At the atomic level, any surface undulation implies the presence of step bunching. Here, the evolution of steps is driven by strain relaxation[2] so that attractive long-range step–step interaction may lead to step bunching instability [8, 9, 32]. With increasing (compressive) strain, the step bunches tend to homogenize, i.e. they are very regular and exhibit a rather narrow distribution of mean distances. This tendency can be exploited to form monodisperse quantum wires of fixed size and spacing (see Sect. 7.6). Such structures are of both fundamental and technological interest, since they can be used as templates for subsequent fabrication of self-organized nanostructures [20, 33].

7.2 Effect of Atomic Steps on the Diffuse Scattering of X-Rays

Steps at interfaces and surface influence the diffuse scattering in the following ways:

[2] Tensile and compressive stress often behave differently. This effect has been investigated but is not yet understood in detail. Initial hints of the origin of the different behaviors of surface steps under compressive and tensile stress are given in [31].

- In most cases the evolution of step bunches results in anisotropic roughness, i.e. the lateral roughness correlation lengths are different for the directions parallel and perpendicular to the surface miscut. Consequently, the diffuse scattering patterns measured parallel to the surface miscut direction and perpendicular to it are different, and the different measured roughness correlation lengths can then be interpreted as the mean terrace sizes in the two directions [13, 34, 35, 36, 37, 38].
- A typical fingerprint of *step-dominated* roughness is an asymmetry in the X-ray diffuse scattering pattern along the miscut direction [13, 35, 36] (see also Sect. 2.6.4).
- The phenomenon of vertical inheritance of interface roughness is well known in amorphous multilayers (e.g. [39, 40]). This cross-correlation of roughness of subsequent interfaces leads to resonant diffuse scattering sheets [13, 35, 39, 40, 41]. However, in the case of stepped interfaces, the direction of inheritance is often inclined with respect to the growth direction [38, 42, 43]. This will lead to a corresponding shearing of the RDS sheets and an asymmetry in the X-ray diffuse scattering profile.
- The morphology of step bunches depends on the growth conditions as well as on the strain incorporated into the step bunches. In the case of very regular step bunches, the interface morphology changes from the self-affine picture of roughness to short-range ordering, and periodic correlation peaks are observed in the diffuse scattering [10, 13, 35, 36, 44].

In the following sections, these phenomena are highlighted and discussed by means of several examples. We shall start with a short discussion about the anisotropic nature of interface roughness in low-strain AlAs/GaAs superlattices and how it depends on the growth parameters of molecular-beam epitaxy (Sect. 7.3). Then we shall evaluate the observed asymmetry in the experimental intensity pattern with reference to the two structural models of (i) inclined inheritance of roughness and (ii) asymmetric vicinal interfaces (Sect. 7.4). Additionally, a simple geometrical model is presented which allows an accurate determination of the inclination angles of roughness inheritance without the need for extensive theoretical simulations based on the DWBA. In Sect. 7.5, a possible application of X-ray diffuse scattering in the GID geometry is discussed that may be able to determine possible atomic disorder in roughness. Finally, in Sect. 7.6, examples of the transition from the self-affine model to rather well-ordered step bunches are presented.

7.3 Anisotropic Interfacial Roughness

We focus on two exemplary samples consisting of 20 period GaAs/AlAs superlattices grown by molecular-beam epitaxy on epiready vicinal GaAs(001). In order to extract the influence of the growth conditions on the interfacial roughness parameters, these samples were chosen to be very similar, i.e. they have nearly the same layer thicknesses and the surface miscut is about 0.40° (with respect to the (001) plane) in the [110] direction. The sample preparation is described in detail in [38]. Briefly, sample 1 was grown in step-flow mode, whereas sample 2 was prepared in two-dimensional nucleation growth mode. The most relevant sample parameters are given in Table 7.1.

Atomic force micrographs highlighting the GaAs surface morphology are displayed in Fig. 7.1. At high magnification (Fig. 7.1a,b), the step structure can be clearly identified for both samples. At low magnification (Fig. 7.1c,d), step bunches can be clearly resolved for sample 1, whereas sample 2 shows 2D islands. Sample 1 exhibits a pronounced anisotropy of the surface morphology that is clearly related to the step structure of the surface. The corresponding terrace lengths are always larger in the $[1\bar{1}0]$ direction. This anisotropy is well known [24] and depends strongly on the surface miscut. Note, however, that the surface miscuts are nearly identical for the two samples under consideration here. In contrast, there is no surface roughness anisotropy visible for sample 2, which was grown in 2D nucleation growth mode.

The anisotropy of the surface morphology of sample 1 can be also observed for the buried interfaces, which can be studied by X-ray diffuse scattering. This can be seen in Fig. 7.2a: the diffuse-scattering intensity patterns near the specularly reflected beam at $q_x = 0$ are different for the directions parallel (AM) and perpendicular (PM) to the surface miscut. In contrast, for 2D nucleation (sample 2), the interface roughness anisotropy vanishes (Fig. 7.2b). For quantitative values, see Table 7.2.

Table 7.1. Structures and growth conditions of the two GaAs/AlAs superlattices grown on GaAs(001) by molecular-beam epitaxy considered here. For more detailed information, see [38]

Sample	1	2
Surface miscut angle (°)	0.38 ± 0.03	0.43 ± 0.03
Growth temperature (°C)	610	580
Growth mode of GaAs layers	Step-flow growth	2D nucleation
Number of periods of AlAs/GaAs	20	20
AlAs layer thickness (nm)	15.60 ± 0.05	16.00 ± 0.05
GaAs layer thickness (nm)	7.10 ± 0.05	6.80 ± 0.05
Total rms roughness (nm)	0.40 ± 0.05	0.35 ± 0.05
Surface rms roughness (nm)	0.42 ± 0.02	0.33 ± 0.02

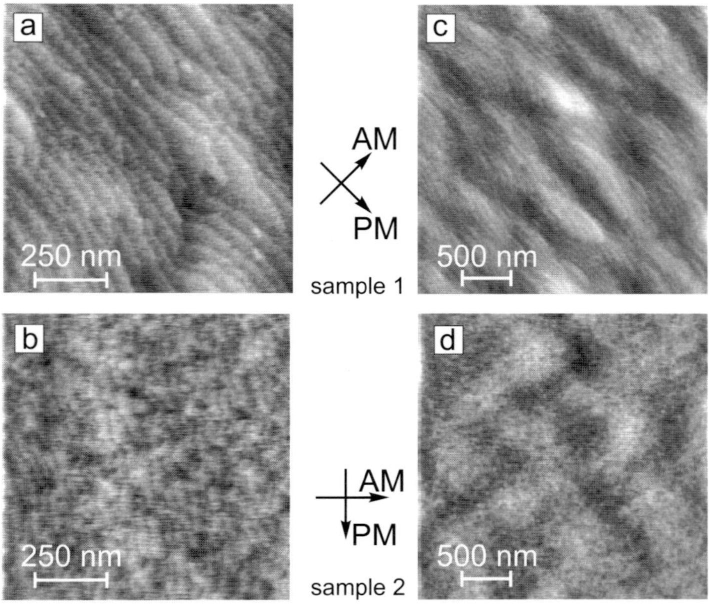

Fig. 7.1. GaAs surface morphology of two GaAs/AlAs superlattices, probed by AFM. The directions parallel (AM) and perpendicular (PM) to the surface miscut are indicated by *arrows*. The AM direction is nearly identical to the [110] direction. This implies that mostly Ga-terminated A-steps along [1$\bar{1}$0] are present, which are known to be more smooth than As-terminated B-steps along [110]. The surface miscut is about 0.40° [38]

Table 7.2. Summary of roughness parameters for the two samples described in Table 7.1. For more detailed information see [38]

Sample	1	2
Along miscut (AM)		
Lateral correlation length $\xi_\parallel$ (nm)	200 ± 25	400 ± 25
Roughness exponent H (Hurst parameter)	0.8 ± 0.1	0.9 ± 0.1
Vertical correlation length $\xi_\perp$ (nm)	150 ± 25	100 ± 25
Angle of inheritance γ_0 (°)	−15 ± 5	45 ± 5
Relative correlated roughness (%)	100	100
Perpendicular to miscut (PM)		
Lateral correlation length $\xi_\parallel$ (nm)	400 ± 25	350 ± 25
Roughness exponent H (Hurst parameter)	0.7 ± 0.1	0.9 ± 0.1
Vertical correlation length $\xi_\perp$ (nm)	300 ± 25	200 ± 25
Angle of inheritance γ_0 (°)	8 ± 5	10 ± 5
Relative correlated roughness (%)	100	100

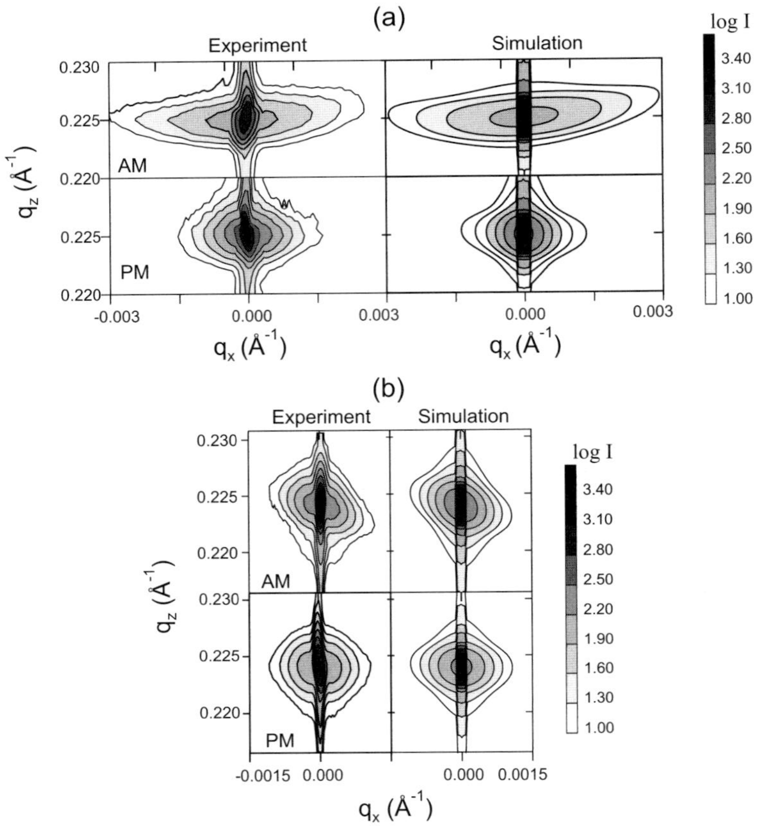

Fig. 7.2. Effect of anisotropy of interfacial roughness in the diffuse scattering (eighth RDS sheet) of two GaAs/AlAs superlattices. (**a**) Sample 1 shows a rather large anisotropy of the interfacial roughness, whereas (**b**) sample 2 shows isotropic roughness. The simulations prove that the inclinations of the RDS sheets are caused by inclined inheritance of interface roughness [38]

7.4 Vertical Inheritance of Interface Roughness

7.4.1 Inclined Inheritance

The lateral roughness parameters of the surface roughness (as measured by AFM) and the interfacial roughness (as measured by X-ray diffuse scattering) are rather similar. This suggests that the roughness parameters of all interfaces are similar and that the roughness is inherited vertically during growth. The vertical inheritance was proven by the observed bunching of the diffuse scattering into resonant diffuse scattering sheets. In Fig. 7.2, the eighth RDS sheet, i.e. the experimental diffuse intensity pattern around the eighth superlattice satellite, is displayed along with corresponding simulations. It is clear from these pictures that the RDS sheet is inclined with respect to the q_x direction, while the center of mass of the RDS sheet is still at $q_x = 0$. This excludes the model of asymmetric vicinal steps (Sect. 2.6.4) and favors the model of inclined inheritance of interface roughness (Sect. 2.6.3).

The theoretical simulations show that the inclination is mainly parallel to the direction of surface miscut (AM); however, there is a significant component of inclined inheritance in the direction perpendicular to the surface miscut (PM), resulting in an effective inheritance in some intermediate direction. Therefore, the direction of inclined inheritance of roughness must be quantified by two angles: the polar angle of inheritance γ_0, which is measured with respect to the surface normal (z axis), and the azimuthal in-plane angle Φ_0, which describes the projection of the direction of inheritance within the interface (Fig. 7.3).

The existence of anisotropic roughness along with inclined inheritance of interface roughness (with unknown angles γ_0 and Φ_0) implies that the intensity distribution of diffuse scattering in reciprocal space has a complicated three-dimensional shape. Therefore, one-dimensional scanning along selected directions, as often performed in earlier work [12, 13, 35, 36, 37, 39, 40, 46], seems unfavorable for a complete characterization of inclined inheritance of roughness. A detailed analysis of roughness and its inheritance can be achieved by use of X-ray scattering simulations that include an appropriate theoretical model for the horizontal and vertical height–height correlation functions. However, in order to correctly evaluate the two angles γ_0 and Φ_0 of inclined inheritance, these simulations must be performed in all three dimensions in reciprocal space. Though it would be extremely useful, this procedure would be very time-consuming and seems not to be feasible even on a fast computer. As an alternative, a simple model is presented below that

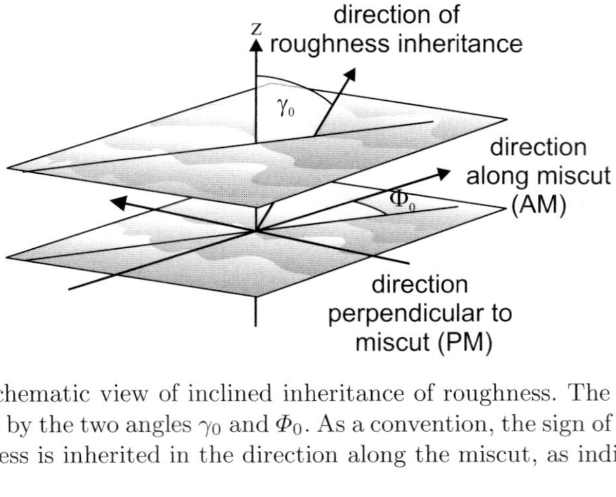

Fig. 7.3. Schematic view of inclined inheritance of roughness. The inclination is characterized by the two angles γ_0 and Φ_0. As a convention, the sign of γ_0 is negative when roughness is inherited in the direction along the miscut, as indicated [45]

enables us to determine the azimuthal and polar angles of inheritance γ_0 and Φ_0, respectively, without the need for extensive 3D theoretical simulations.

7.4.2 Determination of Inclination Angles by a Simple Model

As a consequence of the vertical correlation of the roughness of subsequent interfaces, the diffusely scattered X-ray intensity is bunched into RDS sheets in reciprocal space. According to (2.117), inclined inheritance leads to corresponding shearing of the RDS sheets. A simple model for the evaluation of inclined roughness inheritance in three dimensions has been developed by Schmidbauer et al. [45]. Here, the sheared RDS sheets are modeled by anisotropic sheared ellipsoids. From measurements in different azimuthal sample orientations, the two angles characterizing the inclined inheritance of interface roughness can be determined accurately.

This simple model allows the accurate determination of both the polar angle (γ_0) and the azimuthal angle (Φ_0) of vertical roughness inheritance. In order to describe the anisotropy of the roughness correlation length in different spatial directions, the three-dimensional shape of an RDS sheet is approximated by an ellipsoid with three different principal axes δq_{AM}, δq_{PM}, and δq_z in the directions along the miscut (AM), perpendicular to the miscut (PM), and along the z direction, respectively. These are related to the corresponding lateral and vertical correlation lengths ξ_{AM}, ξ_{PM}, and ξ_z by the expression

$$\delta q_i = c \times 2\pi/\xi_i \quad (i = \mathrm{AM}, \mathrm{PM}, z) \, , \tag{7.1}$$

where the factor c depends on the specific choice of the height–height correlation function used.

In accordance with the shearing of the RDS sheets due to the inclined inheritance[3] of roughness the ellipsoids are sheared along the direction of roughness inheritance. Since the experimental resolution Δq_y in the direction perpendicular to the scattering plane is much larger than the half-width δq_y of the RDS sheet, the diffusely scattered intensity will be almost entirely integrated along the q_y direction. These experimental conditions justify the projection of the sheared ellipsoid onto the q_z–q_x plane.

---

[3] Two different cases have to be considered, and it is very important to distinguish between them, since a wrong assignment of the orientation of the RDS sheet in reciprocal space may lead to a misinterpretation of the data. In case (1), where $\xi_z < \xi_{\mathrm{AM}}$, $\xi_z < \xi_{\mathrm{PM}}$, the RDS sheets are elongated along the q_z direction. The shearing can be then approximated by a rotation around γ_0. As a consequence, the sheared RDS sheets appear to be inclined. In case (2), where $\xi_z > \xi_{\mathrm{AM}}$, $\xi_z > \xi_{\mathrm{PM}}$, the elongation of the ellipsoid is minimum along the q_z direction. Here, the shearing yields rather small angles $\gamma(\Phi)$ (even in the case of large values of γ_0), in the direction opposite to γ_0. Case (1) holds for the samples under investigation here.

As a convention, the direction along the miscut (AM) is defined as $\Phi = 0$ (Fig. 7.3). When the sample is rotated azimuthally around the surface normal (by an angle Φ), different projections of the RDS sheets become visible in the scattering plane (q_x–q_z plane). The projected RDS sheet now appears tilted with respect to the q_z axis by $\gamma(\Phi)$, which can be approximated as

$$\gamma(\Phi) = \arctan\left[\tan\gamma_0 \cos(\Phi - \Phi_0)\right] . \tag{7.2}$$

Figure 7.4a shows the experimental results for the diffuse intensity in the vicinity of an RDS sheet for different values of Φ measured close to the AM direction (top, $\Phi = 5°$–$35°$) and close to the PM direction (bottom, $\Phi = 95°$–$125°$). From these maps, the azimuthal angle of inheritance can be determined with rather high accuracy as $\Phi_0 = 20° \pm 5°$, while the polar angle of inheritance is determined as $\gamma_0 = -20° \pm 7°$. The negative sign of γ_0 indicates that the roughness is inherited in the direction of step flow during growth.

In Fig. 7.4b, corresponding simulations of sheared, azimuthally rotated, projected ellipsoids are shown. Good agreement is found between experiment and simulations, and the best fit to the experimental data yields values of $\gamma_0 = -17° \pm 3°$ and $\Phi_0 = 20° \pm 7°$. There is also excellent agreement when the

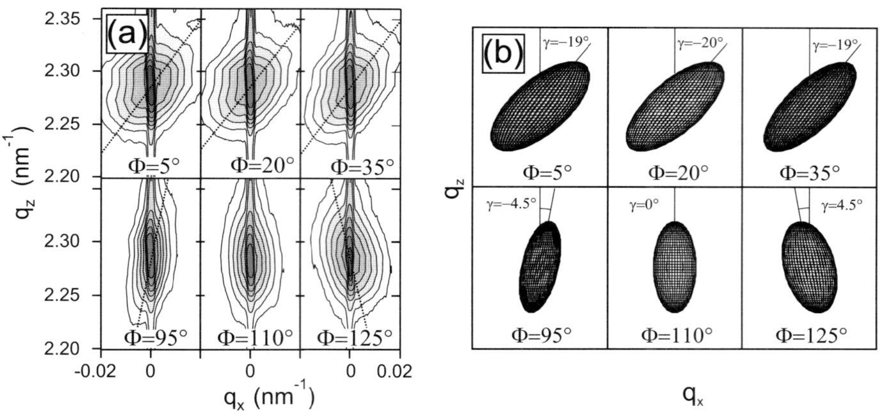

Fig. 7.4. (**a**) Reciprocal-space maps of RDS sheet around the eighth superlattice satellite measured at different azimuths Φ close to the directions parallel (AM, $\Phi = 0°$) and perpendicular (PM, $\Phi = 90°$) to the miscut. The intensity scale varies logarithmically from 10 (*white*) to 2500 (*black*) in arbitrary units. (**b**) Calculated projections of an ellipsoid that is tilted by $\gamma_0 = -17°$ and rotated around the q_z axis by $\Phi = 5°, 20°, 35°$ (*top*) and by $\Phi = 95°, 110°, 125°$ (*bottom*). In order to match the experimental data, the azimuthal angle was chosen as $\Phi_0 = 20°$. The evaluated values of γ are indicated in (**a**) by *dotted lines* for the experimental data and in (**b**) by *solid lines* for the simulations. Note that the scales of the q_x and q_z axes differ by a factor of 2.3 [45]

observed inclination angles γ evaluated from the simulations are compared with the corresponding experimental results (Fig. 7.5).

Our analysis leads to two important consequences. First, the inheritance of roughness approximately follows the direction of step flow during growth. Secondly, the azimuthal direction of inclined inheritance is observed to be 20° off the [110] direction. A possible reason for this offset is the presence of branches of step bunches that do not run along the substrate miscut direction but are azimuthally oriented along the [210] and $[2\bar{1}0]$ directions [47]. The angle between [210] and [110] is about 18°, which is close to the azimuthal angle of inheritance $\Phi_0 = 20°$. Further studies are necessary in order to investigate whether vertical inheritance is indeed influenced by these branches.

7.4.3 Limits of Applicability of Ellipsoidal Model

Finally, we discuss some effects introduced by the finite experimental resolution in different directions in reciprocal space. It has been pointed out above that a detailed analysis of inclined inheritance of roughness requires at least 2D mapping of reciprocal space for a series of values of Φ. As already mentioned, a lot of the work reported in the literature has been performed with single scans along selected directions in reciprocal space. The polar and azimuthal angles of inheritance cannot be determined from such single scans. Moreover, since the RDS sheets have a complex shape in reciprocal space,

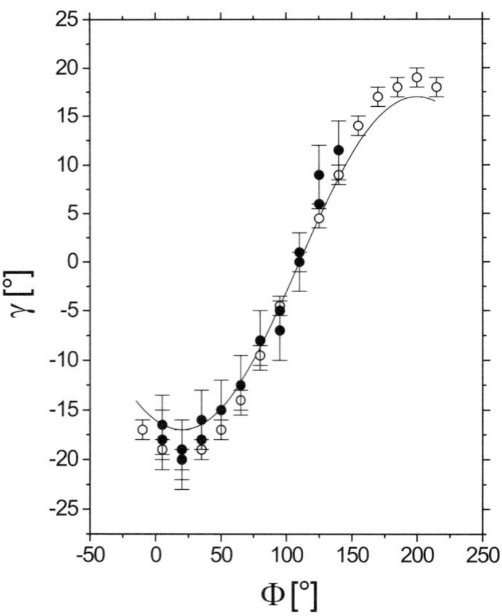

Fig. 7.5. Measured (•) and simulated (∘) values of the angle γ as a function of the azimuthal sample orientation Φ. The simulations were carried out at angles $\gamma_0 = -17°$ and $\Phi_0 = 20°$ ($\Phi = 0°$ corresponds to the direction along the miscut). The *solid line* represents values of γ obtained using (7.2) with $\gamma_0 = -17°, \Phi_0 = 20°$ [45]

the intensity distribution in the single scans is strongly influenced by the experimental resolution in all three directions q_x, q_y, and q_z. In particular, it is found that the vertical resolution Δq_z influences the width of the RDS sheet in the q_x direction as measured by a single transverse scan [38]. Thus, a disregard of the q_z resolution can result in an underestimate of the lateral roughness correlation lengths.

The experiments described above were performed with a laboratory diffractometer which allows a large angular divergence of the incident beam perpendicular to the scattering plane. This large divergence, together with an open detector (in the q_y direction), ensures complete integration of the RDS sheet along q_y. However, with a highly collimated incident beam, as delivered by highly brilliant synchrotron radiation, a sufficiently large value of Δq_y is also achieved, since the resolution in the q_y direction is determined both by the collimation of the incident beam and by the acceptance angle of the scattered beam. The latter is determined by the detector slits and the spot size at the sample in the y direction. Even at a large distance of 1000 mm between the detector and the sample and with an adapted combination of spot size + detector slit in the range of 1 mm, the resulting resolution is of order $\Delta q_y = (2\pi/\lambda) \times \sqrt{2} \times 10^{-3} \approx 0.06\,\mathrm{nm}^{-1}$, which is still larger than the horizontal extension of the RDS sheets considered here. However, for samples with very small roughness correlation lengths, the situation may change. In that case the resolution along q_y would have to be taken into account in the simulation and a modified projection would have to be performed. True two-dimensional slices of reciprocal space can, however, only be recorded by utilizing an additional crystal analyzer behind the sample which is oriented perpendicular to the scattering plane.

The model used here represents a quick and simple alternative to extensive calculations based on the distorted-wave Born approximation using sophisticated height–height correlation functions as discussed in Sect. 2.6. It works nicely in the presence of large lateral anisotropy, which is commonly observed for vicinal semiconductor interfaces, and for ellipsoidally shaped RDS sheets. The corresponding roughness correlation lengths can either be determined by theoretical simulations or be estimated from the vertical and lateral extensions of the RDS sheets measured at different azimuthal sample orientations.

The RDS sheets generated by the interfacial roughness of the AlAs/GaAs superlattice can be approximated by ellipsoids. However, more complicated shapes of resonant diffuse scattering have been reported in the literature:

- RDS sheets that are modulated horizontally owing to lateral positional correlation of step bunches [35, 44],
- S-shaped structures of resonant diffuse scattering due to a highly regular terrace structure which exhibits inclined inheritance [42], and

- different vertical correlations for different spatial frequencies of roughness, which leads to a variation of the vertical width of the RDS sheets depending on the distance from the specular rod [40, 48].

For the first two cases, our simple model seems not to be applicable, and these cases need to be treated by more sophisticated three-dimensional theoretical simulations. The third case is especially important for the interfacial roughness of amorphous multilayers, where a very large range of spatial frequencies is observed. This gives rise to significant diffuse scattering even at very large horizontal momentum transfers, up to $\Delta q_x = 0.5\,\mathrm{nm}^{-1}$. For very flat, stepped semiconductor interfaces ($\sigma = 0.4\,\mathrm{nm}$), the range of spatial frequencies is much smaller. Thus, effects corresponding to this third case are expected to have only a weak influence for the systems considered in the present work.

Despite the limits of our simple approach, the results clearly emphasize that the diffuse scattering from inclined inheritance has to be evaluated in all three dimensions in reciprocal space. In particular, two reciprocal-space maps recorded in perpendicular planes in reciprocal space are generally not sufficient.

7.5 Effect of Roughness in Grazing-Incidence Diffraction

Small-angle X-ray scattering (GISAXS and GIXS) is sensitive to fluctuations in the zeroth Fourier coefficient χ_0 of the dielectric polarizability $\chi(\boldsymbol{r})$. On the other hand, lattice strains are commonly probed by X-ray diffraction. Here, both of the fluctuations $\delta\chi_0$ and $\delta\chi_h$ contribute to the diffuse scattering. Interface sensitivity is provided by using a scattering geometry with grazing incidence and exit (grazing-incidence diffraction, GID). A possible application of X-ray diffuse scattering in GID is to determine the atomic ordering in the roughness. If the atomic lattice is not distorted at the interfaces, the parameters given by GIXS should fit the GID data as well. If, however, there is lattice disorder in the interface morphology, then χ_h is reduced by some static Debye–Waller factor and the diffuse scattering measured in GID will be lower than one would expect from the GIXS measurements.

In Fig. 7.6, the integrated rocking curve in the vicinity of the 220 reciprocal-lattice point is displayed along with theoretical simulations that have been convoluted with the experimental resolution function [49]. Note that the simulations were carried out for the coherent part of the X-ray diffraction and do not include diffuse scattering. The experiment demonstrates the high quality of the superlattice, since the half-widths of the Bragg peak and the superlattice peaks correspond to the calculated parameters. Moreover, the intensity of the specular beam exhibits a maximum at the first superlattice peak, thus proving that the diffraction is dynamical. The

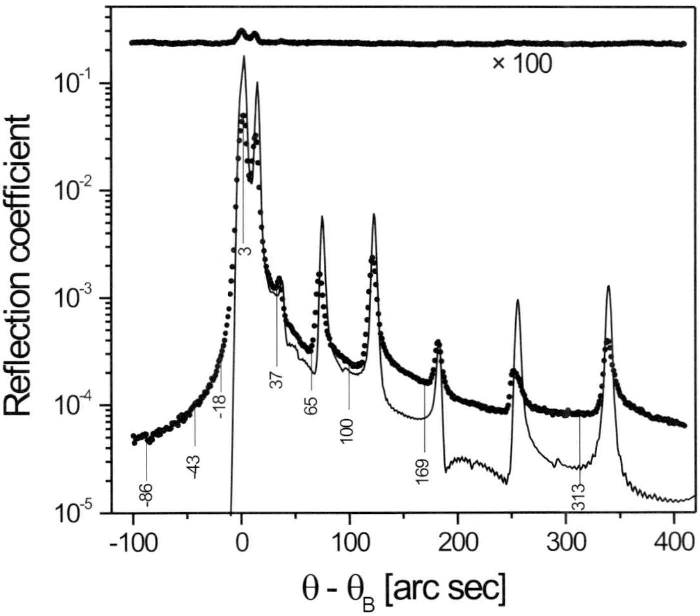

Fig. 7.6. Integrated intensity from an AlAs/GaAs superlattice in the vicinity of 220 measured in GID geometry at $\alpha_i = 0.5°$. *Dots* represent experimental data, and the *solid line* is a theoretical simulation convoluted with the experimental resolution function. The *top line* is the magnified reflection coefficient for the specular reflected beam. *Vertical marks* indicate the angular positions of the sample where PSD spectra of the diffracted intensity were recorded (see Fig. 7.7) [49]

intensity in the wings of the experimental GID curve is noticeably higher than the theoretical simulations. This discrepancy can be attributed to diffuse scattering. In particular, this holds for the low-angle side of the rocking curve, which exclusively shows diffuse scattering, while the coherently diffracted part must vanish owing to total *internal* reflection of the diffracted wave.

The diffuse scattering, measured as a function of the exit angle at various values of θ (as marked in Fig. 7.6), is displayed in Fig. 7.7. The experimental data may be compared with corresponding simulations. The total intensity is composed of

1. the coherent Bragg peak (marked by an arrow),
2. diffuse scattering due to surface roughness, and
3. diffuse scattering from interface roughness.

The diffuse scattering due to interface roughness forms regular superlattice peaks, with a maximum intensity around the position of the coherent peak. These superlattice peaks are sections through RDS sheets that are

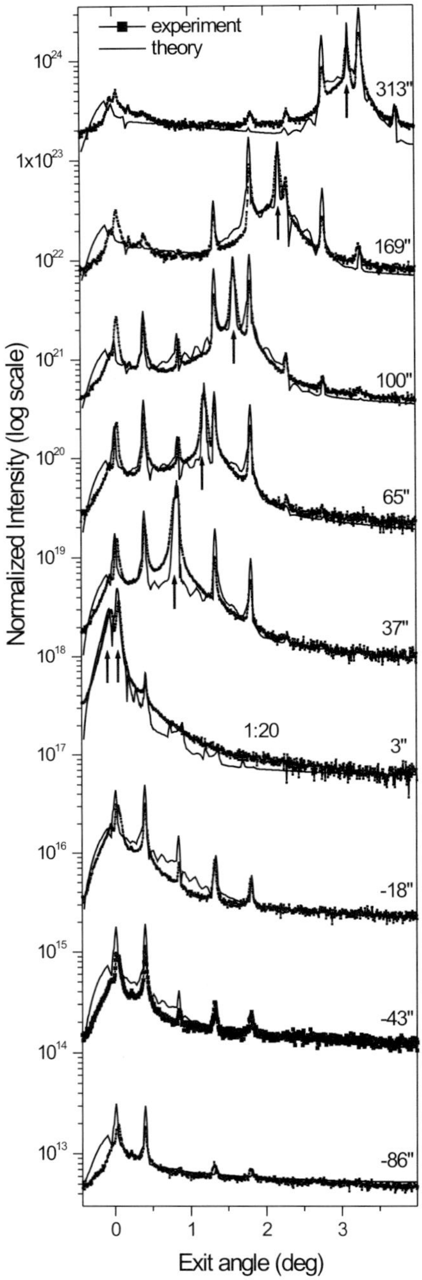

Fig. 7.7. Diffuse intensity of GID as a function of the exit angle at different values of $\theta - \theta_B$ (θ_B is the Bragg angle) for an AlAs/GaAs superlattice. *Dots* represent experimental data, and *solid lines* are calculations with parameters taken from corresponding studies of GIXS. *Arrows* mark coherent peaks [49]

also visible in GIXS. It is striking that the RDS sheets are also present at $\Delta\theta = \theta - \theta_B < 0$, although the coherent signal vanishes here. This proves that the X-ray wave field in GID has a larger penetration into the superlattice, even though the diffracted wave is trapped at the surface.

The simulations shown in Fig. 7.7 were performed with the same parameters as used for the GIXS investigations. The overall satisfactory agreement between experiment and simulations indicates that the atomic structure is not distorted at the interfaces.

The surface roughness leads to a maximum in the diffuse scattering near the critical angle of total external reflection ($\alpha_c = 0.31°$). The intensity of diffuse scattering from surface roughness is too large in the simulations. Better agreement can be achieved if the surface roughness is reduced from $\sigma = 9$ Å (as found in GIXS investigations) to $\sigma = 4.5$ Å. The difference in the observed σ-values is due to the distinction between diffuse scattering in GIXS and GID. The crystal structure at the surface could have been partially destroyed by oxidation, giving rise to weaker diffuse scattering in GID. Thus, the experiment confirms the conclusion that the measurement of diffuse scattering in GID may provide information about the atomic ordering at interfaces which is not accessible by GIXS.

7.6 Ordered Linear Structures

So far, the surface and interface morphology of rather strongly disordered systems have been considered. This is related to the original meaning of the term *roughness*, in that it implies a statistical distribution of the height fluctuations at the surface or interfaces. To some extent, a self-affine interface or surface undergoes random up and down fluctuations. There are no dominating spatial frequencies, and their statistical distribution fulfills certain scaling relations. It has already been mentioned that the self-affine model cannot be valid for an infinite range of spatial frequencies. Therefore, a cutoff length, called the roughness correlation length, was introduced. This correlation length describes an upper limit for the length scale up to which the self-affine description is valid. Since the cutoff is rather soft (it is applied by introducing an exponential decay of the height–height correlation function), the diffuse scattering exhibits a rather smooth decay of intensity as a function of the scattering vector $\boldsymbol{q}$.

On the other hand, the self-affine model also needs some modifications at low length scales, at least because of the presence of atoms or the discrete distribution of atomic steps and corresponding terraces. This leads to a distinct asymmetry in the diffuse scattering; however, the intensity distribution is still smooth as can be seen in Figs. 7.2 and 7.4.

The situation changes when the interface morphology consists of a regular array of well-ordered steps. In this case a distinct spatial frequency dominates, and corresponding correlation *peaks* show up in the diffuse scattering. An example is presented in Fig. 7.8, where the surface of a five-period $In_{0.19}Ga_{0.81}As/GaAs$ superlattice grown on vicinal GaAs is displayed. A surface miscut of $\alpha = 2.5°$ in the direction of B[1$\bar{1}$0] was determined by HRXRD, and the layer thicknesses are $d_{GaAs} = 4.2\,nm$ and $d_{InGaAs} = 2.5\,nm$. The surface morphology reveals regular step bunches running along the B[110] direction, with a mean spacing of about 150–200 nm and typical heights of 4–6 nm, these heights being of the same order as the superlattice period $t = d_{InGaAs} + d_{GaAs} = 6.7\,nm$.

The corresponding X-ray diffuse scattering in the vicinity of the symmetrical 004 reciprocal-lattice point (in the HRXRD geometry) is displayed in Fig. 7.9. Perpendicular to the miscut (Fig. 7.9a), only the substrate peak with a surrounding cloud of diffuse scattering and the vertically modulated crystal truncation rod can be resolved. In contrast, a complicated pattern of diffuse scattering is visible in the direction of the miscut (Fig. 7.9b). A schematic view of this pattern is shown in Fig. 7.9c, which indicates the main features that are visible in the diffuse scattering:

- The vertical stripes (rods) are due to lateral positional correlation of step bunches on internal interfaces. These rods are inclined with respect to the CTR, and they are oriented exactly parallel to the [001] direction. This asymmetry with respect to the mean surface (which is represented by the CTR) has already been discussed in Sect. 2.6.4. The exact orientation of

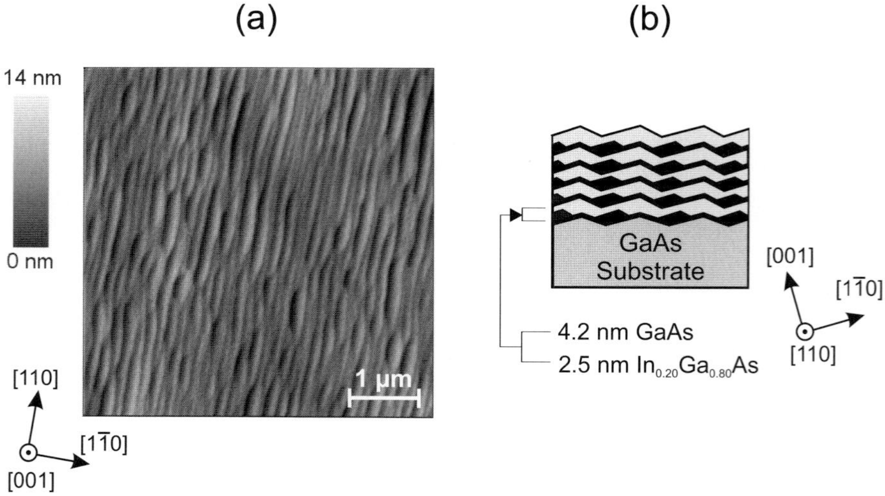

Fig. 7.8. (**a**) Atomic force micrograph of the GaAs surface of a five-period $In_{0.19}Ga_{0.81}As/GaAs$ superlattice grown on vicinal GaAs. (**b**) Schematic view of the superlattice geometry. The (001) GaAs substrate has a miscut of 2.5° in the direction of B[1$\bar{1}$0]). The sample was grown by metal-organic chemical vapor deposition (MOCVD)

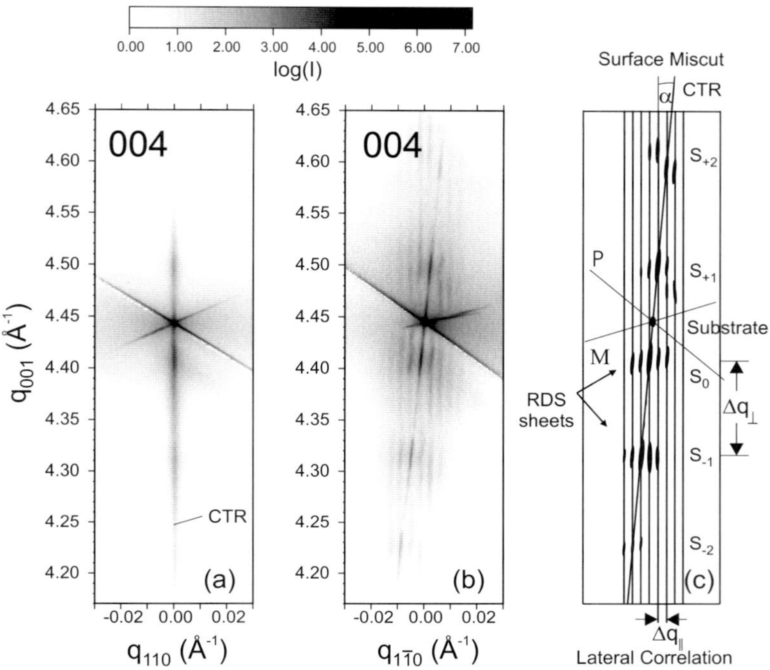

Fig. 7.9. X-ray diffuse scattering from a five-period $In_{0.19}Ga_{0.81}As$/GaAs superlattice grown on vicinal GaAs in the vicinity of the 004 reciprocal-lattice point: (**a**) perpendicular to miscut, (**b**) parallel to miscut. (**c**) represents a schematic drawing of (**b**)

the satellite rods parallel to q_{001} implies that lateral positional correlation exactly parallel to the crystallographic direction $[1\bar{1}0]$ is present.

- Although they are chopped up by the lateral correlation rods, there are pronounced RDS sheets visible in the intensity pattern, indicating vertical correlation.
- Surprisingly, the CTR always crosses the zero-order lateral correlation peak of the RDS sheet. This means that the miscut angle α fulfills the interesting relationship $\sin\alpha = \Delta q_{\|}/\Delta q_{\perp} = d_{\perp}/\langle d_{\|}\rangle$. As a consequence of this, the height of the step bunches should be approximately equal to the superlattice period $d_{\perp}$.

The experimental results obtained by X-ray diffuse scattering are summarized in Fig. 7.10a, and they qualitatively agree with results obtained by transmission electron microscopy (Fig. 7.10b). However, the TEM data clearly indicate that the GaAs substrate does not show any initial step bunching. This means that the step bunches are not a consequence of replicating the morphology of a faceted, grooved GaAs surface, but are generated during the growth process only. Their occurrence can presumably be ascribed to strain

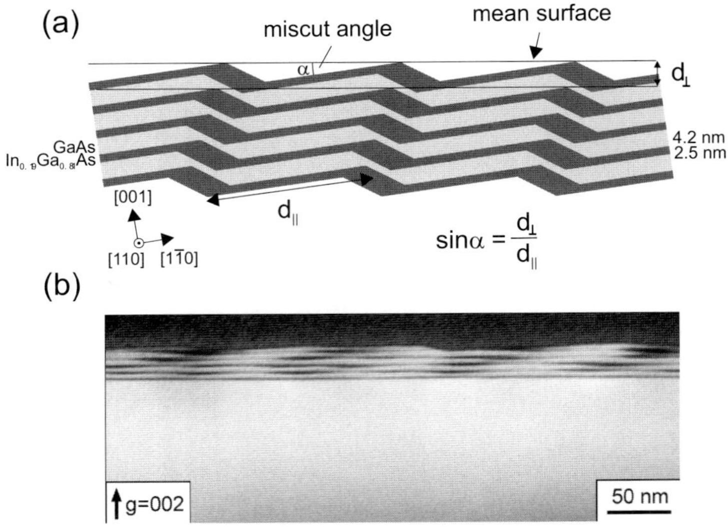

Fig. 7.10. Structural model of superlattice: (**a**) schematic view reconstructed from the X-ray diffuse scattering pattern. The relationship $\sin\alpha = \Delta q_\parallel/\Delta q_\perp = d_\perp/\langle d_\parallel\rangle$ implies that the height of the step bunches is approximately given by the superlattice period $d_\perp$. (**b**) Cross-sectional TEM (by courtesy of H. Kirmse, Humboldt University Berlin)

relaxation during growth. From the TEM micrograph, it is also obvious that the height of the step bunches increases slowly with increasing layer index. The first layer closest to the substrate shows only a small step bunch height, and this height increases as growth progresses. Nevertheless, the existence of RDS sheets in the X-ray diffuse intensity pattern clearly indicates vertical correlation of the step bunches in the different layers. The RDS sheets appear to be inclined, indicating inclined inheritance. This seems to be confirmed by the TEM data in Fig. 7.10b.

The existence of high-order correlation peaks offers the possibility of testing the various SRO and LRO models introduced in Sect. 2.3.4. In Fig. 7.11a, a horizontal section parallel to $q_{1\bar{1}0}$ through the first RDS sheet is presented, along with a line shape analysis using Gaussian profiles. The satellite peaks are equidistantly spaced, yielding a mean lateral spacing of $\langle d_\parallel\rangle = 1671\,\text{Å}$. The standard deviation of the satellite peaks, $\delta q_{1\bar{1}0}$, increases with the satellite peak index. This is shown in Fig. 7.11c. The *random-walk* SRO model given by (2.57) and (2.58), which predicts a quadratic increase of the peak width with increasing peak index, does not exactly fit the experimental data. However, using this model, the rms standard deviation of the mean lateral spacing $\langle d_\parallel\rangle$ can be estimated as $\sigma = 120$–$140\,\text{Å}$, which approximately fits the value $\sigma = 151\,\text{Å}$ determined directly from the autocorrelation function $C(r)$ in real space (Fig. 7.11b), which is the inverse Fourier transform of

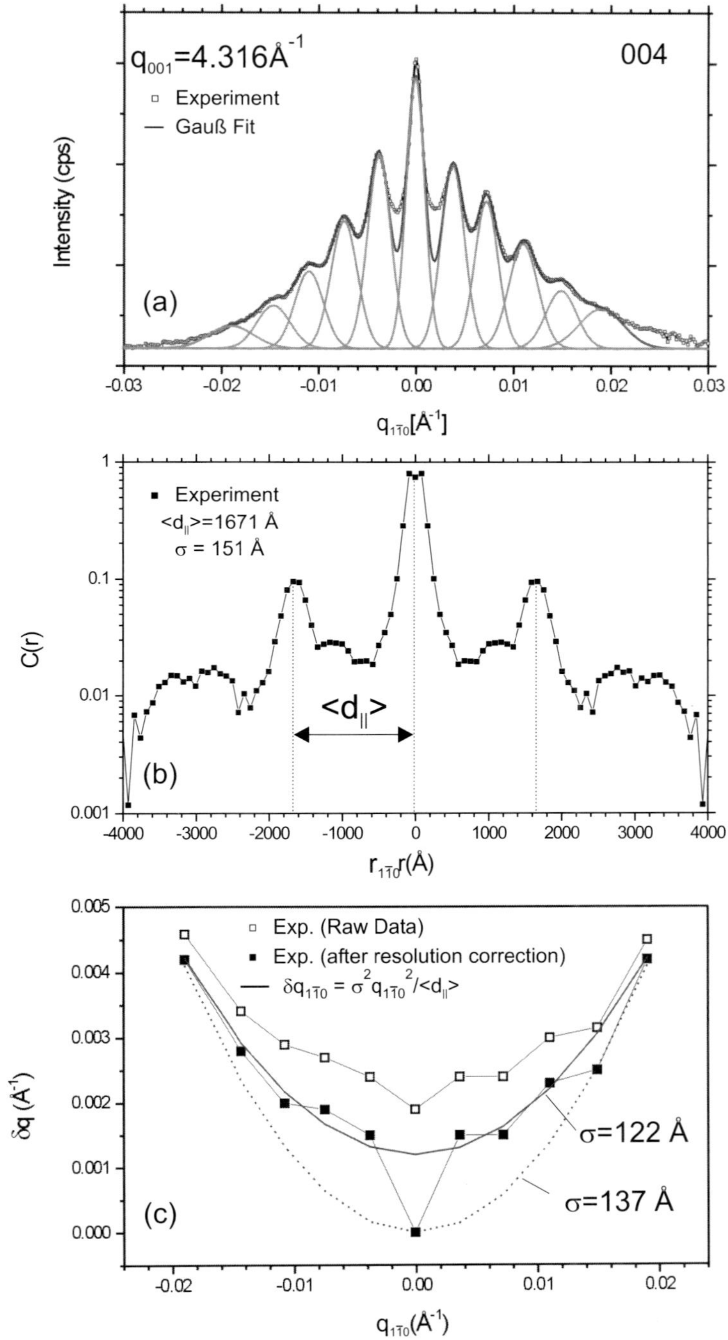

Fig. 7.11. (**a**) Horizontal section through the first RDS sheet of Fig. 7.9b at $q_{001} = 4.316\,\text{Å}^{-1}$, (**b**) the corresponding autocorrelation function $C(r)$ in real space, (**c**) experimental peak widths (FWHM) as a function of $q_{1\bar{1}0}$ and corresponding calculations using (2.58)

the experimental intensity distribution (see (2.55)). The strong damping of higher orders in $C(r)$ clearly indicates SRO of the step bunches; however, this damping indicates a correlation length of $\xi = 10\,\mu\text{m}$, which seems far too high.

References

1. E. Runge, J. Menniger, U. Jahn, R. Hey, and H. T. Grahn, Phys. Rev. B **52**, 12207 (1995).
2. U. Jahn, J. Menniger, S. Kwok, E. Runge, F. Fujiwara, R. Hey, and H. Grahn, Phys. Stat. Sol. (b) **150**, 439 (1995).
3. D. W. Pohl, W. Denk, and M. Lanz, Appl. Phys. Lett. **44**, 651 (1984).
4. E. Betzig and J. Trautmann, Science **257**, 189 (1992).
5. K. Brunner, G. Abstreiter, G. Böhm, G. Tränkle, and G. Weimann, Phys. Rev. Lett. **73**, 1138 (1994).
6. S. K. Sinha, E. B. Sirota, S. Garoff, and H. B. Stanley, Phys. Rev. B **38**, 2297 (1988).
7. A.-L. Barabasi and H. E. Stanley, *Fractal Concepts in Surface Growth* (Cambridge University Press, Cambridge, 1995).
8. J. Tersoff, Y. H. Phang, Z. Zhang, and M. G. Lagally, Phys. Rev. Lett. **75**, 2730 (1995).
9. F. Liu, J. Tersoff, and M. G. Lagally, Phys. Rev. Lett. **80**, 1268 (1998).
10. T. A. Rabedeau, I. M. Tidswell, P. S. Pershan, J. Bevk, and B. S. Freer, Appl. Phys. Lett. **59**, 3422 (1991).
11. Y. H. Phang, D. E. Savage, T. F. Kuech, M. G. Lagally, J. S. Park, and K. L. Wang, Appl. Phys. Lett. **60**, 2986 (1992).
12. Z. H. Ming, A. Krol, Y. L. Soo, Y. H. Kao, J. S. Park, and K. L. Wang, Phys. Rev. B **47**, 16373 (1993).
13. R. L. Headrick and J.-M. Baribeau, Phys. Rev. B **48**, 9174 (1993).
14. V. Holý and T. Baumbach, Phys. Rev. B **49**, 10668 (1994).
15. M. Lässig, Phys. Rev. Lett. **77**, 526 (1996).
16. K. Sudoh, T. Yoshinobu, H. Iwasaki, and E. D. Williams, Phys. Rev. Lett. **80**, 5152 (1998).
17. S. Song and S. G. J. Mochrie, Phys. Rev. B **51**, 10068 (1995).
18. S. Song, S. G. J. Mochrie, and G. B. Stephenson, Phys. Rev. Lett. **74**, 5240 (1995).
19. S. Song, M. Yoon, S. G. J. Mochrie, G. B. Stephenson, and S. T. Milner, Surf. Sci. **372**, 37 (1997).
20. J. Zhu, K. Brunner, G. Abstreiter, O. Kienzle, F. Ernst, and M. Rühle, Phys. Rev. B **60**, 10935 (1999).
21. R. Nötzel, U. Jahn, Z. Niu, A. Trampert, J. Fricke, H. Schönherr, T. Kurth, D. Heitmann, L. Däweritz, and K. H. Ploog, Appl. Phys. Lett. **72**, 2002 (1998).
22. T. Kato, T. Takeuchi, Y. Inoue, S. Hasegawa, K. Inoue, and H. Nakashima, Appl. Phys. Lett. **72** (1998).
23. R. L. Schwoebel and E. J. Shipsey, J. Appl. Phys. **37**, 3682 (1966).
24. R. Hey, M. Wassermeier, J. Behrend, L. Däweritz, K. Ploog, and H. Raidt, J. Cryst. Growth **154**, 1 (1995).

25. C. Orme, M. D. Johnson, K.-T. Leung, B. G. Orr, P. Smilauer, and D. Vvedensky, J. Cryst. Growth **150**, 128 (1995).
26. M. D. Johnson, C. Orme, A. W. Hunt, D. Graff, J. Sudijono, L. M. Sander, and B. G. Orr, Phys. Rev. Lett. **72**, 116 (1994).
27. Z. M. Wang, L. Däweritz, and K. H. Ploog, Surf. Sci. **459**, L482 (2000).
28. R. J. Asaro and W. A. Tiller, Metall. Trans. **3**, 1789 (1972).
29. M. A. Grinfel'd, Sov. Phys. Dokl. **31**, 831 (1986).
30. D. J. Srolovitz, Acta Metall. **37**, 621 (1989).
31. V. M. Kaganer and K. H. Ploog, Phys. Rev. B **64**, 205301 (2001).
32. G. Patriarche, A. Ougazzaden, and F. Glas, Appl. Phys. Lett. **76**, 306 (2000).
33. J. Stangl, T. Roch, V. Holý, M. Pinczolits, G. Springholz, G. Bauer, I. Kegel, T. H. Metzger, J. Zhu, K. Brunner, et al., J. Vac. Sci. Technol. B **18**, 2187 (2000).
34. S. K. Sinha, M. K. Sanyal, S. K. Satija, C. F. Majkrzak, D. A. Neumann, H. Homma, S. Szpala, A. Gibeaud, and H. Morkoc, Physica B **198**, 72 (1994).
35. Y. H. Phang, C. Teichert, M. G. Lagally, L. J. Peticolas, J. C. Bean, and E. Kasper, Phys. Rev. B **50**, 14435 (1994).
36. R. L. Headrick, J.-M. Baribeau, and Y. E. Strausser, Appl. Phys. Lett. **66**, 96 (1995).
37. J. Stettner, L. Schwalowsky, O. H. Seeck, M. Tolan, W. Press, C. Schwarz, and H. von Känel, Phys. Rev. B **53**, 1398 (1996).
38. E. A. Kondrashkina, S. A. Stepanov, R. Opitz, M. Schmidbauer, R. Köhler, R. Hey, M. Wassermeier, and D. V. Novikov, Phys. Rev. B **56**, 10469 (1997).
39. T. Salditt, T. Metzger, and J. Peisl, Phys. Rev. Lett. **73**, 2228 (1994).
40. E. Spiller, D. Stearns, and M. Krumrey, J. Appl. Phys. **74**, 107 (1993).
41. D. G. Stearns, J. Appl. Phys. **71**, 4286 (1992).
42. J. Stangl, V. Holý, J. Grim, G. Bauer, J. Zhu, K. Brunner, G. Abstreiter, O. Kienzle, and F. Ernst, Thin Solid Films **357**, 71 (1999).
43. C. Giannini, T. Baumbach, D. Lübbert, R. Felici, L. Tapfer, T. Marschner, W. Stolz, N. Y. Jin-Phillipp, and F. Phillipp, Phys. Rev. B **61**, 2173 (2000).
44. C. Giannini, L. Tapfer, Y. Zhuang, L. D. Caro, T. Marschner, and W. Stolz, Phys. Rev. B **55**, 5276 (1997).
45. M. Schmidbauer, R. Opitz, T. Wiebach, and R. Köhler, Phys. Rev. B **64**, 195316 (2001).
46. Y. H. Phang, R. Kariotis, D. E. Savage, and M. G. Lagally, J. Appl. Phys. **72**, 4627 (1992).
47. K. Hata, H. Shigekawa, T. Okano, T. Ueda, and M. Akiyama, Phys. Rev. B **55**, 7039 (1997).
48. R. Paniago, H. Homma, P. C. Chow, S. C. Moss, Z. Barnea, S. S. P. Parkin, and D. Cookson, Phys. Rev. B Rapid Commun. **52**, 17052 (1995).
49. S. A. Stepanov, E. A. Kondrashkina, M. Schmidbauer, R. Köhler, J. U. Pfeiffer, T. Jach, and A. Y. Souvorov, Phys. Rev. B **54**, 8150 (1996).

A Appendix

A.1 The Resolution Function in Reciprocal Space

A.1.1 General Expression

By using the relationships

$$
\begin{aligned}
q_x &= k\left(\cos\alpha_\mathrm{f}\ \cos\theta_\mathrm{f} - \cos\alpha_\mathrm{i}\cos\theta_\mathrm{i}\right) , \\
q_y &= k\left(\cos\alpha_\mathrm{f}\ \sin\theta_\mathrm{f} + \cos\alpha_\mathrm{i}\ \sin\theta_\mathrm{i}\right) , \\
q_z &= k\left(\sin\alpha_\mathrm{f} + \sin\alpha_\mathrm{i}\right) ,
\end{aligned} \tag{A.1}
$$

and the general formula for error propagation for a function $f(x_1, \ldots, x_n)$ with n variables x_i and corresponding uncertainties Δx_1,

$$
\Delta f = \sum_i \left|\frac{\partial f}{\partial x_i}\right| \Delta x_i \ , \tag{A.2}
$$

we find

$$
\begin{aligned}
\Delta q_x &= k\,\big(\sin\alpha_\mathrm{f}\cos\theta_\mathrm{f}\,\Delta\alpha_\mathrm{f} + \cos\alpha_\mathrm{f}\sin\theta_\mathrm{f}\,\Delta\theta_\mathrm{f} + \sin\alpha_\mathrm{i}\cos\theta_\mathrm{i}\,\Delta\alpha_\mathrm{i} \\
&\qquad + \cos\alpha_\mathrm{i}\sin\theta_\mathrm{i}\,\Delta\theta_\mathrm{i}\big) + q_x\frac{\Delta k}{k} \ ,
\end{aligned} \tag{A.3a}
$$

$$
\begin{aligned}
\Delta q_y &= k\,\big(\sin\alpha_\mathrm{f}\sin\theta_\mathrm{f}\,\Delta\alpha_\mathrm{f} + \cos\alpha_\mathrm{f}\cos\theta_\mathrm{f}\,\Delta\theta_\mathrm{f} + \sin\alpha_\mathrm{i}\sin\theta_\mathrm{i}\,\Delta\alpha_\mathrm{i} \\
&\qquad + \cos\alpha_\mathrm{i}\cos\theta_\mathrm{i}\,\Delta\theta_\mathrm{i}\big) + q_y\frac{\Delta k}{k} \ ,
\end{aligned} \tag{A.3b}
$$

$$
\Delta q_z = k\,\big(\cos\alpha_\mathrm{f}\,\Delta\alpha_\mathrm{f} + \cos\alpha_\mathrm{i}\,\Delta\alpha_\mathrm{i}\big) + q_z\frac{\Delta k}{k} \ . \tag{A.3c}
$$

A.1.2 High-Resolution X-Ray Diffraction

For coplanar diffraction and neglecting dispersion of the wave vector, we can set $\theta_\mathrm{i} = \theta_\mathrm{f} = 0$ and obtain

$$
\Delta\boldsymbol{q}^{\mathrm{HRXRD}} = k \begin{pmatrix} \sin\alpha_\mathrm{i}\,\Delta\alpha_\mathrm{i} + \sin\alpha_\mathrm{f}\,\Delta\alpha_\mathrm{f} \\ \cos\alpha_\mathrm{i}\,\Delta\theta_\mathrm{i} + \cos\alpha_\mathrm{f}\,\Delta\theta_\mathrm{f} \\ \cos\alpha_\mathrm{i}\,\Delta\alpha_\mathrm{i} + \cos\alpha_\mathrm{f}\,\Delta\alpha_\mathrm{f} \end{pmatrix} . \tag{A.4}
$$

Note that although only coplanar scattering is treated here, the resolution in the q_y direction has been included. This is due to our generalized definition of *coplanar* (see Sect. 3.6.1).

The experiment can be performed by using either

1. an area detector, or
2. a linear PSD along with a slit perpendicular to the direction resolved by the detector.

In the first case, both of the divergences $\Delta\theta_{\mathrm{i}}$ and $\Delta\alpha_{\mathrm{i}}$ have to be chosen small. At the same time, the detector acceptance angles $\Delta\theta_{\mathrm{f}}$ and $\Delta\alpha_{\mathrm{f}}$ also have to be reduced by using a rather small beam size in both spatial directions. In this case, all three components $\Delta q_{x,y,z}$ are of roughly equal magnitude (see Table 3.1), which is very advantageous concerning 3D reciprocal-space mapping. In the second case, the detector acceptance angle $\Delta\theta_{\mathrm{f}}$ is rather large, so that the corresponding divergence of the incident beam $\Delta\theta_{\mathrm{i}}$ can also be chosen rather large. However, this affects only the resolution in the q_y direction, whereas the resolutions in the q_x and q_z directions remain unchanged. The setup with a PSD along with a slit is thus most advantageous for 2D reciprocal-space mapping, whereby the intensity is integrated over a significant distance Δq_y (Table 3.1).

A.1.3 Grazing-Incidence Diffraction

In the case of grazing-incidence diffraction, the angles of incidence and exit $\alpha_{\mathrm{i,f}}$ are small, so that we can write $\sin\alpha_{\mathrm{i,f}} \approx \alpha_{\mathrm{i,f}}$ and $\cos\alpha_{\mathrm{i,f}} \approx 1$. This reduces the resolution function to

$$\Delta \boldsymbol{q}^{\mathrm{GID}} = k \begin{pmatrix} \sin\theta_{\mathrm{i}}\,\Delta\theta_{\mathrm{i}} + \sin\theta_{\mathrm{f}}\,\Delta\theta_{\mathrm{f}} \\ \cos\theta_{\mathrm{i}}\,\Delta\theta_{\mathrm{i}} + \cos\theta_{\mathrm{f}}\,\Delta\theta_{\mathrm{f}} \\ \Delta\alpha_{\mathrm{i}} + \Delta\alpha_{\mathrm{f}} \end{pmatrix} . \tag{A.5}$$

Owing to the grazing-incidence geometry and the large scattering angles, the lateral size of the X-ray beam after leaving the sample is rather large. This circumstance impedes the application of an area detector, which is useful only when both the vertical and the lateral beam sizes are very small. Therefore, the θ_{f} resolution has to be ensured by either

1. using a triple-crystal arrangement with a crystal analyzer (Fig. 3.6) behind the sample, or, when only medium resolution is required,
2. using a collimating detector slit system.

In both cases, the out-of-plane resolution Δq_z is limited mainly by $\Delta\alpha_{\mathrm{i}}$, which is the horizontal divergence of the synchrotron beam (Table 3.1). On the other hand, the in-plane resolution is determined by the divergences $\Delta\theta_{\mathrm{i,f}}$. In the first case these are very small, resulting in excellent in-plane resolution Δq_x and Δq_y. In the case where a collimating detector slit system is used, the in-plane resolution is deteriorated by one order of magnitude (Table 3.1).

A.1.4 Grazing-Incidence Small-Angle X-Ray Scattering

In the GISAXS geometry *all angles* are small, and we can write $\sin(\alpha_{\mathrm{i,f}}, \theta_{\mathrm{i,f}}) \approx (\alpha_{\mathrm{i,f}}, \theta_{\mathrm{i,f}})$ and $\cos(\alpha_{\mathrm{i,f}}, \theta_{\mathrm{i,f}}) \approx 1$. The corresponding resolution can then be expressed as

$$\Delta \boldsymbol{q}^{\mathrm{GISAXS}} = k \begin{pmatrix} \alpha_{\mathrm{i}}\, \Delta\alpha_{\mathrm{i}} + \alpha_{\mathrm{f}}\, \Delta\alpha_{\mathrm{f}} + \theta_{\mathrm{i}}\, \Delta\theta_{\mathrm{i}} + \theta_{\mathrm{f}}\, \Delta\theta_{\mathrm{f}} \\ \Delta\theta_{\mathrm{i}} + \Delta\theta_{\mathrm{f}} \\ \Delta\alpha_{\mathrm{i}} + \Delta\alpha_{\mathrm{f}} \end{pmatrix} . \tag{A.6}$$

We may distinguish between three different scattering geometries using

1. an area detector (CCD),
2. a vertically aligned PSD together with a crystal analyzer, and
3. a horizontally aligned PSD.

Independent of these different geometries, the resolution in the q_x direction is approximately one order of magnitude better than that along q_y. However, as pointed out before, this excellent resolution is associated with a serious restriction on the accessible range of q_x. Another feature of the GISAXS geometry is that the resolution is practically constant in the q_y and q_z directions, whereas Δq_x is a function of all four angles. Typical values of the resolution in all three geometries are summarized in Table 3.1.

A.2 Strain Analysis by the Finite-Element Method

Nowadays, the finite-element method is an established technique which is widely applied to solve complex problems in physics and engineering science. Among these problems are the linear and nonlinear treatment of stress and strain in solids, and the calculation of temperature gradients and thermal-conduction fields. In addition, it is applied to treat the mechanics of fluids. Here, we focus on strain analysis based on linear elasticity theory. For a deeper insight, a large choice of textbooks about finite-element analysis is available (e.g. [1, 2, 3]).

A.2.1 A Very Brief Survey of Linear Elasticity Theory

First, we briefly review those parts of linear elasticity theory which are essential for a basic understanding. For a more detailed study, several textbooks are available, e.g. [4, 5].

In the continuum approach, the deformation of a strained atomic lattice is described by a vector field $\boldsymbol{u}(\boldsymbol{r})$ which quantifies the displacement field with respect to the ideal undeformed lattice $\boldsymbol{r}$. Therefore, the positions of the atoms in the strained lattice can be written as

$$\boldsymbol{r}' = \boldsymbol{r} + \boldsymbol{u}(\boldsymbol{r}) \ . \tag{A.7}$$

It is advantageous to treat relative changes of deformation, and, for small displacements, to consider only the first derivative $\partial u_i/\partial x_j$ of $\boldsymbol{u}$. This tensor can, in general, be split into a totally symmetric part $\varepsilon_{ij}(\boldsymbol{r})$ and a totally antisymmetric part $\omega_{ij}(\boldsymbol{r})$:

$$\varepsilon_{ij} = \frac{1}{2}\left(\frac{\partial u_i}{\partial x_j} + \frac{\partial u_j}{\partial x_i}\right) , \tag{A.8a}$$

$$\omega_{ij} = \frac{1}{2}\left(\frac{\partial u_i}{\partial x_j} - \frac{\partial u_j}{\partial x_i}\right) . \tag{A.8b}$$

The antisymmetric part describes rigid rotations of an element of the lattice, whereas the symmetric strain tensor ε_{ij} describes deformations of an element of the lattice. These deformations can be classified in terms of the local changes of the lattice spacings (described by the diagonal elements of ε_{ij}) and in terms of lattice shearing (described by the nondiagonal elements of ε_{ij}). The diagonal elements (often called the *normal strain*) generally induce a change of the volume, whereas shearing is a volume-invariant transformation. Owing to its intrinsic symmetry properties, the strain tensor contains six independent parameters.

The basic assumption of linear elasticity theory is the validity of Hooke's law, which interconnects the components of the stress tensor $\sigma_{ij}(\boldsymbol{r})$ and the strain tensor $\varepsilon_{ij}(\boldsymbol{r})$ via a linear relationship (Einstein's convention of summation over double identical indices is applied throughout):

$$\sigma_{ij} = C_{ijkl}\varepsilon_{kl} \ . \tag{A.9}$$

This law goes back to Robert Hooke, who, on the basis of experiments with springs, stretched wires, and coils, stated a linear rule between extension and force. Hooke first announced the law 1676 in the form of an anagram he placed at the end of an article, "ceiinossittuu", which he revealed could be deciphered as "ut tensio, sic uis", meaning "as the extension, so the force". He published this law in 1678 [6]. Hooke's law is equivalent to the assumption of *harmonic* potentials between neighboring atoms in an atomic lattice, which is valid at small values of ε_{ij}.

The stress tensor σ_{ij} includes the forces that induce the deformation. By definition, σ_{ij} is the ith component of the force per unit area on a plane whose outward-drawn normal is parallel to the positive x_j direction. Owing to the symmetry of σ_{ij} and ε_{kl}, Hooke's tensor C_{ijkl} is also symmetric, and has 36 independent parameters. Therefore, the elastic coefficients C_{ijkl} are often represented in a contracted 6×6 matrix notation as C_{mn}, where m and n are indices corresponding to a pair of indices ij or kl, e.g. $C_{11} := C_{1111}$, $C_{12} := C_{1122}$, $C_{44} := C_{2323}$.

Since the work of deformation is a state variable, the elements of C_{mn} are symmetrical, and the number of independent parameters is reduced to 21. Crystal symmetry can further decrease this number. For an isotropic medium, there are just two independent parameters, e.g. the compression modulus K and the shear modulus μ. Alternatively, the Lamé constants λ and μ (where μ is again the shear modulus) are often used. For cubic symmetry, three independent parameters remain. These are the elastic constants C_{11}, C_{12}, and C_{44}, which are tabulated in Table A.1. The elastic anisotropy can be quantified by the anisotropy coefficient

$$S = \frac{C_{11} - C_{12}}{2C_{44}} , \tag{A.10}$$

which takes typical values of around $S = 0.5$ for cubic semiconductors, to be compared with the isotropic value of $S = 1$.

Note the difference between *elastic* strain and *total* strain. The difference becomes important when materials with different lattice parameters are considered. When one is talking about *elastic strain*, the reference lattice is defined by the intrinsic nondeformed ideal lattice of the material concerned. In contrast, the term *total strain* is used when a common reference lattice is used for both materials. Hooke's law is, of course, valid only for the elastic strain. However, after the strain tensor components have been calculated, it can easily be transformed into the total strain by adding a constant offset. The use of total strain is most advantageous for X-ray simulations, since the atomic displacement is described with respect to the ideal lattice of the substrate.

Table A.1. Lattice parameter a, elastic constants C_{mn}, and elastic anisotropy coefficient S of selected semiconductor materials [7, 8, 9][a]

Material	a (Å)	C_{11} (GPa)	C_{12} (GPa)	C_{44} (GPa)	S
Si	5.431	166	64.0	79.6	0.64
Ge	5.658	126	44.0	67.7	0.61
GaAs	5.653	119	53.4	59.6	0.55
InAs	6.058	83.4	45.4	39.5	0.48
InP	5.869	101.1	56.1	45.6	0.49
GaP	5.451	140.5	62.0	70.3	0.56

[a] A comprehensive database of semiconductors and materials parameters can be found on the website of the Ioffe Institute, St. Petersburg [9].

A.2.2 Equilibrium Conditions

The starting point of the finite-element method for strain analysis is the total elastic strain energy

$$\Pi(\boldsymbol{u}(\boldsymbol{r})) = \frac{1}{2}\int_{\mathrm{V}} \sigma_{ij}(\boldsymbol{r})\varepsilon_{ij}(\boldsymbol{r})\,\mathrm{d}V - \int_{\mathrm{V}} \boldsymbol{f}\cdot\boldsymbol{u}\,\mathrm{d}V - \int_{\mathrm{S}} \boldsymbol{T}\cdot\boldsymbol{u}\,\mathrm{d}S\,, \tag{A.11}$$

which has to be minimized as a function of the strain field $\boldsymbol{u}(\boldsymbol{r})$ with consideration of the boundary conditions. Here $\boldsymbol{f}$ and $\boldsymbol{T}$ are the external volume and surface forces, respectively. The minimum can be found by applying the *principle of virtual displacements*, which states that – under the condition of fulfillment of the boundary conditions – the internal virtual work associated with virtual displacements $\delta\boldsymbol{u}$ equals the total external virtual work. When no external forces are present and if (A.9) is used, this is equivalent to

$$\delta\Pi = \int_{\mathrm{V}} \delta\varepsilon_{ij}\, C_{ijkl}\varepsilon_{kl}\,\mathrm{d}V = 0\,. \tag{A.12}$$

A.2.3 The Finite-Element Method

The finite-element method represents a numerical procedure to solve (A.12). The steps of the procedure can be divided as follows:

1. The system under investigation is divided into discrete three-dimensional subelements ("FEM cells") that are interconnected by nodes.
2. A system of algebraic equations is created using (A.12), whereby the boundary conditions are taken into account.
3. The system of equations is solved iteratively and the corresponding nodal displacements $\boldsymbol{u}(\boldsymbol{r})$ are calculated.
4. Finally, all other state variables (e.g. the strain tensor ε_{ij} and the elastic strain energy density ϵ) can be calculated.

In Fig. A.1, an exemplary FEM model that has been used for the calculation of a SiGe island attached to a Si(001) substrate is displayed. After creation of the model, the boundary conditions of the limiting cells of the substrate have to be defined. In the case displayed here, they can be formulated as follows:

1. The lowest layer of cells in the substrate is fixed in all three spatial directions: $u(x, y, z) = 0$.
2. The cells in the other outer limiting areas of the substrate may move only within the respective area: $u_x(x, y, z) = 0$ for the y–z plane and $u_y(x, y, z) = 0$ for the x–z plane.

The material constants that are used in the FEM are the elastic constants C_{mn} and the lattice parameters of the species considered (Table A.1). For a binary compound, both can be calculated by assuming Vegard's law [10]:

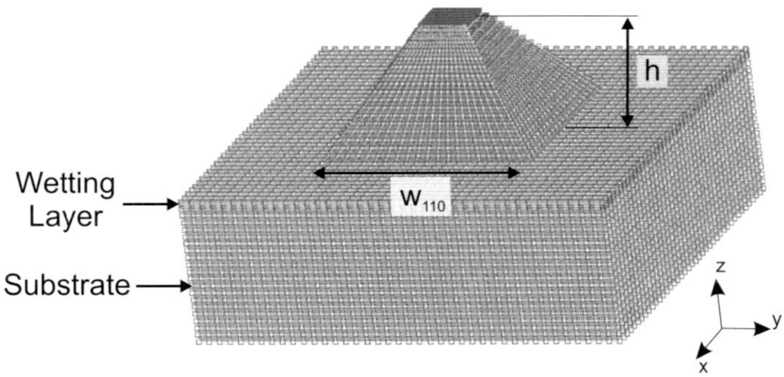

Fig. A.1. Three-dimensional FEM model of a SiGe truncated pyramid (base width w_{110}, height h) attached to a Si(001) substrate. The FEM elements are interconnected by *white lines*, which cross each other at the FEM nodes

$$C_{mn}^{\mathrm{Si}_{1-x}\mathrm{Ge}_x} = xC_{mn}^{\mathrm{Ge}} + (1-x)C_{mn}^{\mathrm{Si}} \,, \tag{A.13a}$$

$$a^{\mathrm{Si}_{1-x}\mathrm{Ge}_x} = xa^{\mathrm{Ge}} + (1-x)a^{\mathrm{Si}} \,. \tag{A.13b}$$

The basic assumption of Vegard's empirical rule is that the atomic volumes of the constituent elements are independent of the alloy concentration, regardless of the local lattice distortion. Recent research [11] has, however, shown that there are significant deviations from Vegard's law. For SiGe, deviations up to 6% have been found [12]. Therefore, a quadratic relationship between x and the lattice mismatch is often used rather than a simple linear behavior.

A.2.4 FEM as Input for X-Ray Scattering Simulations

The deformation field obtained by the FEM cannot be used directly in the simulations of X-ray diffuse scattering described in Sect. 2.3.3. There are two main reasons for this:

1. Although cuboids are used, the FEM mesh is generally not regular.
2. The FEM mesh has rather large cell sizes in the range of $a = 50$ Å. These will lead to artifacts in the X-ray diffuse scattering pattern, since the corresponding minimum size in reciprocal space is given by $\Delta q = 2\pi/a \approx 0.1$ Å^{-1}, which is the regime where diffuse scattering is investigated.

Therefore, the FEM mesh has to be refined and regularized before it can be used as an input for X-ray diffuse-scattering simulations. This is done with a linear approximation. With fast computers, the regularization procedure can be performed rapidly even if cell sizes in the range of a few atomic spacings are used. Typically, the total number of cells is $N = 10^5$–10^6, although this varies from case to case.

A.3 List of Acronyms

Table A.2. List of acronyms used in this book

Symbol	Meaning
2D	two-dimensional
3D	three-dimensional
AFM	atomic force microscopy
AM	direction along miscut
CCD	charge-coupled device
CL	cathodoluminescence
CTR	crystal truncation rod
DFT	density functional theory
DWBA	distorted-wave Born approximation
EDXS	energy-dispersive X-ray spectroscopy
EELS	electron energy loss spectroscopy
ESRF	European Synchrotron Radiation Facility
FEM	finite-element method
FWHM	full width at half maximum
GID	grazing-incidence diffraction
GISAXS	grazing-incidence small-angle X-ray scattering
GIXS	grazing-incidence X-ray scattering
HASYLAB	Hamburger Synchrotronstrahlung-Labor
HRXRD	high-resolution X-ray diffraction
KMC	kinetic Monte Carlo simulation
LPE	liquid-phase epitaxy
LRO	long-range order
MBE	molecular-beam epitaxy
MOCVD	metal-organic chemical vapor deposition
PL	photoluminescence
PM	direction perpendicular to miscut
PSD	position-sensitive detector
RDS	resonant diffuse scattering
SAXS	small-angle X-ray scattering
SEM	scanning electron microscopy
SNOM	scanning near-field optical microscopy
SRO	short-range order
TEM	transmission electron microscopy
VFF	valence force field
XSR	X-ray specular reflectometry

A.4 List of Symbols

Table A.3. List of symbols used in this book

Symbol	Meaning
A	molar atomic mass
$A(\boldsymbol{q})$	amplitude of scattered wave
a	lattice parameter
$C(\boldsymbol{R})$	height–height correlation function
$C(\boldsymbol{r})$	autocorrelation function
C_{ijkl}	elastic coefficients
C_{mn}	elastic coefficients in reduced representation
$C_{nn'}(\boldsymbol{r})$	height–height cross-correlation function
c	island area coverage
c_0	vacuum light velocity
$\boldsymbol{D}(\boldsymbol{r},t)$	electric induction vector
d	layer thickness
d	step height
$\langle d \rangle$	mean lateral spacing of mesoscopic structures
$\mathrm{d}\sigma/\mathrm{d}\Omega$	differential scattering cross section
$\boldsymbol{E}(\boldsymbol{r},t)$	electric field vector
e	electron charge
$F_{\boldsymbol{h}}$	structure amplitude
f	lattice mismatch
$f'(\omega), f''(\omega)$	Hönl corrections
f_0	atomic form amplitude
$G(\boldsymbol{q})$	interference function
$G(\boldsymbol{r},\boldsymbol{r}')$	Green's function
$g(\boldsymbol{R})$	height-difference correlation function
H	Hurst parameter
h	island height
$\boldsymbol{h}$	reciprocal-lattice vector
h_{c}	critical thickness
K	deflection parameter
$\boldsymbol{k}^{\mathrm{f}}$	wave vector of scattered beam
$\boldsymbol{k}^{\mathrm{i}}$	wave vector of incoming beam
L	coherence length
M	transfer matrix
m	electron mass
N_{A}	Avogadro's number
n_0	complex refractive index
P	polarization factor
$P(q_x,q_y)$	in-plane power spectrum
p	polarization anisotropy

Table A.3. (continued)

Symbol	Meaning
Q	geometric aspect ratio (of an island)
$\boldsymbol{q}$	scattering vector
$\boldsymbol{R}$	distance vector
r_0	classical electron radius
r_F	Fresnel reflection coefficient
S	elastic anisotropy
t	superlattice period
t_F	Fresnel transmission coefficient
$\boldsymbol{u}(\boldsymbol{r})$	displacement vector
$\hat{\boldsymbol{V}}$	scattering-potential operator
V_c	critical island size
V_u	volume of elementary unit cell
w	island base width
x	chemical composition
$\langle\alpha\rangle$	strength of asymmetry (of strain)
α_c	critical angle of total external reflection
α_f	angle of scattered beam with respect to sample surface
α_i	angle of primary beam with respect to sample surface
β	imaginary part of deviation of refractive index from unity
γ	ratio of electron energy to rest mass
γ_0	polar angle of inheritance
δ	real part of deviation of refractive index from unity
ϵ	elastic strain energy density
ε_0	vacuum permittivity
ε_{ij}	elastic strain tensor
2θ	scattering vector
θ_B	Bragg angle
θ_f	in-plane angle of scattered beam
θ_i	in-plane angle of incident beam
$\varLambda$	X-ray penetration depth
λ	X-ray wavelength
λ_u	period length of undulator
μ	linear absorption coefficient
ξ	correlation length
$\varrho(\boldsymbol{r})$	electron density
$\varrho_\mathrm{A}(\boldsymbol{r})$	atomic electron density
$\varrho_\mathrm{m}(\boldsymbol{r})$	mass density
σ	surface/interface roughness
$\varPhi_0$	azimuthal angle of inheritance
$\chi(\boldsymbol{r},\omega)$	dielectric polarizability
$\varOmega(\boldsymbol{r})$	(island) shape function
$\varOmega^\mathrm{FT}(\boldsymbol{q})$	(island) form factor

A.5 Physical Constants

Table A.4. Useful fundamental physical constants

Symbol	Meaning	Value
c_0	vacuum light velocity	$2.998 \times 10^8 \mathrm{m\,s^{-1}}$
e	electron charge	1.602×10^{-19} C
h	Planck's constant	6.626×10^{-34} J s
m	electron mass	9.109×10^{-31} kg
N_A	Avogadro's number	6.022×10^{26} molecules $\mathrm{kmole^{-1}}$
r_0	classical electron radius	2.818×10^{-15} m
ε_0	vacuum permittivity	$8.854 \times 10^{-12} \mathrm{C^2\,N^{-1}\,m^{-1}}$

References

1. K.-J. Bathe, *Finite Element Procedures in Engineering Analysis* (Prentice Hall, Englewood Cliffs, NJ, 1982).
2. F. L. Matthews, G. A. O. Davies, D. Hitchings, and C. Soutis, *Finite Element Modelling of Composite Materials and Structures* (CRC Press, Boca Raton, 2001).
3. T. J. R. Hughes, *The Finite Element Method: Linear Static and Dynamic Finite Element Analysis* (Dover, New York, 2000).
4. J. P. Hirth and J. Lothe, *Theory of Dislocations* (Krieger, Malabar, FL, 1992).
5. L. D. Landau and E. M. Lifshitz, *Theory of Elasticity* (Pergamon, Oxford, 1986), 3rd ed.
6. R. Hooke, Lectures *"De Potentia Restitutiva", or of Spring. Explaining the Power of Springing Bodies.* British Museum, London (Copy) p. 23 (1678).
7. O. Madelung, ed., *Semiconductors – Group IV Elements and III–V Compounds* (Springer, Berlin, Heidelberg, 1991).
8. S. R. M. Levinstein and M. Shur, eds., *Handbook Series on Semiconductor Parameters*, vol. 1, 2 (World Scientific, London, 1996, 1999).
9. Ioffe Institute, St. Petersburg, www.ioffe.rssi.ru.
10. L. Vegard, Z. Phys. **5**, 17 (1921).
11. G. Ice and C. Sparks, Annu. Rev. Mater. Sci. **29**, 25 (1999).
12. J. P. Dismukes, L. Ekstrom, and R. J. Paff, J. Phys. Chem **68**, 10 (1964).

Index

Springer Tracts in Modern Physics

162 **Electroweak Precision Tests at LEP**
By W. Hollik and G. Duckeck 2000. 60 figs. VIII, 161 pages

163 **Symmetries in Intermediate and High Energy Physics**
Ed. by A. Faessler, T.S. Kosmas, and G.K. Leontaris 2000. 96 figs. XVI, 316 pages

164 **Pattern Formation in Granular Materials**
By G.H. Ristow 2000. 83 figs. XIII, 161 pages

165 **Path Integral Quantization and Stochastic Quantization**
By M. Masujima 2000. 0 figs. XII, 282 pages

166 **Probing the Quantum Vacuum**
Pertubative Effective Action Approach in Quantum Electrodynamics and its Application
By W. Dittrich and H. Gies 2000. 16 figs. XI, 241 pages

167 **Photoelectric Properties and Applications of Low-Mobility Semiconductors**
By R. Könenkamp 2000. 57 figs. VIII, 100 pages

168 **Deep Inelastic Positron-Proton Scattering in the High-Momentum-Transfer Regime of HERA**
By U.F. Katz 2000. 96 figs. VIII, 237 pages

169 **Semiconductor Cavity Quantum Electrodynamics**
By Y. Yamamoto, T. Tassone, H. Cao 2000. 67 figs. VIII, 154 pages

170 **d–d Excitations in Transition-Metal Oxides**
A Spin-Polarized Electron Energy-Loss Spectroscopy (SPEELS) Study
By B. Fromme 2001. 53 figs. XII, 143 pages

171 **High-T_c Superconductors for Magnet and Energy Technology**
By B. R. Lehndorff 2001. 139 figs. XII, 209 pages

172 **Dissipative Quantum Chaos and Decoherence**
By D. Braun 2001. 22 figs. XI, 132 pages

173 **Quantum Information**
An Introduction to Basic Theoretical Concepts and Experiments
By G. Alber, T. Beth, M. Horodecki, P. Horodecki, R. Horodecki, M. Rötteler, H. Weinfurter, R. Werner, and A. Zeilinger 2001. 60 figs. XI, 216 pages

174 **Superconductor/Semiconductor Junctions**
By Thomas Schäpers 2001. 91 figs. IX, 145 pages

175 **Ion-Induced Electron Emission from Crystalline Solids**
By Hiroshi Kudo 2002. 85 figs. IX, 161 pages

176 **Infrared Spectroscopy of Molecular Clusters**
An Introduction to Intermolecular Forces
By Martina Havenith 2002. 33 figs. VIII, 120 pages

177 **Applied Asymptotic Expansions in Momenta and Masses**
By Vladimir A. Smirnov 2002. 52 figs. IX, 263 pages

178 **Capillary Surfaces**
Shape – Stability – Dynamics, in Particular Under Weightlessness
By Dieter Langbein 2002. 182 figs. XVIII, 364 pages

179 **Anomalous X-ray Scattering for Materials Characterization**
Atomic-Scale Structure Determination
By Yoshio Waseda 2002. 132 figs. XIV, 214 pages

180 **Coverings of Discrete Quasiperiodic Sets**
Theory and Applications to Quasicrystals
Edited by P. Kramer and Z. Papadopolos 2002. 128 figs., XIV, 274 pages

Springer Tracts in Modern Physics

181 **Emulsion Science**
Basic Principles. An Overview
By J. Bibette, F. Leal-Calderon, V. Schmitt, and P. Poulin 2002. 50 figs., IX, 140 pages

182 **Transmission Electron Microscopy of Semiconductor Nanostructures**
An Analysis of Composition and Strain State
By A. Rosenauer 2003. 136 figs., XII, 238 pages

183 **Transverse Patterns in Nonlinear Optical Resonators**
By K. Staliūnas, V. J. Sánchez-Morcillo 2003. 132 figs., XII, 226 pages

184 **Statistical Physics and Economics**
Concepts, Tools and Applications
By M. Schulz 2003. 54 figs., XII, 244 pages

185 **Electronic Defect States in Alkali Halides**
Effects of Interaction with Molecular Ions
By V. Dierolf 2003. 80 figs., XII, 196 pages

186 **Electron-Beam Interactions with Solids**
Application of the Monte Carlo Method to Electron Scattering Problems
By M. Dapor 2003. 27 figs., X, 110 pages

187 **High-Field Transport in Semiconductor Superlattices**
By K. Leo 2003. 164 figs.,XIV, 240 pages

188 **Transverse Pattern Formation in Photorefractive Optics**
By C. Denz, M. Schwab, and C. Weilnau 2003. 143 figs., XVIII, 331 pages

189 **Spatio-Temporal Dynamics and Quantum Fluctuations in Semiconductor Lasers**
By O. Hess, E. Gehrig 2003. 91 figs., XIV, 232 pages

190 **Neutrino Mass**
Edited by G. Altarelli, K. Winter 2003. 118 figs., XII, 248 pages

191 **Spin-orbit Coupling Effects in Two-dimensional Electron and Hole Systems**
By R. Winkler 2003. 66 figs., XII, 224 pages

192 **Electronic Quantum Transport in Mesoscopic Semiconductor Structures**
By T. Ihn 2003. 90 figs., XII, 280 pages

193 **Spinning Particles – Semiclassics and Spectral Statistics**
By S. Keppeler 2003. 15 figs., X, 190 pages

194 **Light Emitting Silicon for Microphotonics**
By S. Ossicini, L. Pavesi, and F. Priolo 2003. 206 figs., XII, 284 pages

195 **Uncovering *CP* Violation**
Experimental Clarification in the Neutral K Meson and B Meson Systems
By K. Kleinknecht 2003. 67 figs., XII, 144 pages

196 **Ising-type Antiferromagnets**
Model Systems in Statistical Physics and in the Magnetism of Exchange Bias
By C. Binek 2003. 52 figs., X, 120 pages

197 **Electroweak Processes in External Electromagnetic Fields**
By A. Kuznetsov and N. Mikheev 2003. 24 figs., XII, 136 pages

198 **Electroweak Symmetry Breaking**
The Bottom-Up Approach
By W. Kilian 2003. 25 figs., X, 128 pages

199 **X-Ray Diffuse Scattering from Self-Organized Mesoscopic Semiconductor Structures**
By M. Schmidbauer 2004. 100 figs., X, 204 pages

Printing: Saladruck Berlin
Binding Lüderitz&Bauer, Berlin